Science Makes the World Go Round

Michael Böcher · Max Krott

Science Makes the World Go Round

Successful Scientific Knowledge Transfer for the Environment

Springer

Michael Böcher
Forest and Nature Conservation Policy
University of Göttingen
Göttingen
Germany

Max Krott
Forest and Nature Conservation Policy
University of Göttingen
Göttingen
Germany

English language editing by Eduardo Fargas.

ISBN 978-3-319-34077-7 ISBN 978-3-319-34079-1 (eBook)
DOI 10.1007/978-3-319-34079-1

Library of Congress Control Number: 2016939362

Illustrations by Masumi Akiyama: Figs. 1.1, 1.2, 2.1, 2.2, all factsheets and success factor lists in Chap. 3, Figs. 5.1, 5.2, 5.3, 5.4, 5.5, 5.6, 5.7.

Printed on acid-free paper

This Springer imprint is published by Springer Nature
The registered company is Springer International Publishing AG Switzerland

To Christian Smoliner,
the most innovative Chief Science
Administrator in Vienna, Austria

Preface

There is a consensus that, when scientific ideas are put into practice, they should contribute to the improvement of environmental solutions. However, although the transfer of scientific knowledge into practice has been the object of many activities, results are poor, and seldom has better science led to better policy.

Three decades ago, we approached the issue of knowledge transfer by consulting and evaluating the Austrian Cultural Landscape Programme, which was one of the first in Europe to conduct transdisciplinary research in an effort to make a difference in the practice of the sustainable use of landscapes. We learned that most of the expectations scientists had about political actors were not met, and that, conversely, political actors did not acquire the information they expected from science. Nevertheless, a few success stories showed that a valuable scientific contribution to practical solutions is not impossible. We analysed the success stories based on political science theory and identified various success factors. By advising the German Federal Environment Agency, one of the key players in Europe in environmental policy support, we had the chance to develop our model further and to generate the RIU (research, integration and utilization) model, which provided the basis for this book.

Today, the RIU model has proved to be useful in the identification of the key factors involved in the transfer of scientific knowledge in European and international cases. Consideration of these factors allows the development of a road map for the professionalization of the scientific support for policy. The RIU model is not an attempt to diminish the importance of numerous innovative institutions and activities involved in the transfer of scientific knowledge from theory to practice, or to substitute them. On the contrary, in that the RIU model presents a professional approach to the identification of key factors involved in knowledge transfer within the spectrum of existing institutions and activities. One hopes that the ability to focus on said factors might improve the effectiveness of those efforts, which are growing rapidly in number. This could even lead to a new profession, one of "integrators": people who build bridges successfully between researchers and political actors.

We want to thank all scientists who have trusted us and given us deep insights into their daily business of research, and into the—often frustrating—efforts to make science relevant for practice. We also thank the political actors who shared with us their experiences in the struggle for solutions for environmental issues. Special thanks also go to the editors at Springer, who were enthusiastic about our topic and who made possible the production of this book.

Göttingen
February 2016

Michael Böcher
Max Krott

Acknowledgments

The research results presented in this book originate from several research projects that were conducted during the last couple of years, but especially from "Optimierung des Wissenstransfers aus den Umweltwissenschaften", which was funded by the Austrian Federal Ministry of Science and Research (BMWF) from 2009 to 2013, as part of its research programme, proVISION—provision for Nature and Society, and from "Umsetzung des Konzepts einer modernen Ressortforschung im Geschäftsbereich des BMU", which was funded from 2009 to 2010 by the German Federal Agency for the Environment with financial means from the Federal Ministry of the Environment, Nature Conservation and Nuclear Safety.

We wish to thank all interviewees for providing us with deep insight into their daily business as it covered research, integration and utilization. We also thank our colleagues for helping us improve our RIU model of scientific knowledge transfer through their criticism and discussion, especially the working group for scientific knowledge transfer at the Chair of Forest and Nature Conservation Policy at the Georg-August-University of Göttingen, Germany, namely Budi Dharmawan, Janina Heim, Do Thi Huong and Kenji Nagasaka.

We also thank Masumi Akiyama for creating the illustrations in this book and Eduardo Fargas for his exceptional work in English language editing.

Contents

List of Figures

List of Tables

Chapter 1
Making Science Work in Policy and Politics

But Professor, what are you thinking of? No one has yet become popular by being right. Only when being right suits some political party can that happen.
—Arthur Schnitzler (1993; first: 1912)

1.1 Joint Solving of Problems by Policy and Science

The magic promise of policy and science is to improve life for mankind. The basic idea is to handle and solve problems actively instead of accepting the threats and hardship of life. In doing so, science and policy face the great challenges of the world, like feeding people, protecting nature, keeping peace or maintaining health. There is no doubt that such problems are too great to solve for either science or policy by itself, and that the only strategy that can make a difference is for both areas of knowledge to join their efforts.

Science strives to deepen our insight into the complex causes and consequences of problems. Scientific modelling is able to show how climate will change in the future and how temperature, wind and rain will influence forests and agriculture. Based on particular scenarios, political actors better understand the long and short-term consequences of climate change and gather information about the potential usefulness of new political instruments and their effects. Whereas science helps to achieve in-depth understanding of a subject, the everyday life experience and knowledge of an informed layperson are not sufficient to understand world problems and solve them effectively.

Political actors depend on scientific information to design adequate policies, and science alone cannot effectively improve life without relying on strong policy. Political actors only have the mandate and the ability to intervene in societies and economies in order to put science-based solutions into practice. The need for joint action by politics and science is overwhelming; both claim to be aware of the other's potential and of the benefits that their cooperation would bring to both.

Despite the willingness to cooperate, the examples of successful science-based political advice and policy are rare. Scientific knowledge does not flow easily into

M. Böcher and M. Krott, *Science Makes the World Go Round*,
DOI 10.1007/978-3-319-34079-1_1

political practice. Scientists and political actors are often not even able to communicate on a specific issue. Sometimes it would almost seem that they come from different planets and speak different languages. Our aim with this book is to form a clear picture of these differences and to look for possible bridges to improve joint problem solving in practice even despite existing communication problems.

1.2 Sharing a Beer with a Politician

Imagine what would happen if a scientist met an old friend who has become a politician. They would sit down to have a beer and exchange the experiences they have had with the effects of climate change on society. The researcher might talk about his new outstanding climate model. The model describes multiple dimensions of climate change combining mean temperature with maxima and minima over time. "This is very nice", the politician says, "but tell me one thing: Will the climate become warmer or colder in the next 10 years?" "It depends," will be the scientist's answer, "it depends on what you focus on, the maxima or the mean temperature". "What a pity", the politician thinks, "My colleagues and I cannot attend a seminar to become experts in climate change definitions. What we need is a straightforward message to get people interested and involved."

The scientist continues his enthusiastic report: "Yesterday I got some new, exciting results showing that temperature has been cooling down throughout the last decade. Such data suggests that my entire model has to be reformulated. We are expecting an improved model soon." "Wow", the politician thinks, "how crazy scientific progress is—models have to be abolished for innovation, while in politics we need to convince people of our arguments and stay with these arguments once they have been agreed on in order to have decisions be approved and to keep them that way."

"My dear friend, tell me where I can get good scientific advice on how to manage forests in order to create an optimal carbon sink." "A scientific answer is not yet available, because the growth rates of tree species under conditions of climate change are not yet sufficiently understood," the scientist replies. "Good scientific evidence will be available in the future only." "My dear scientist", the politician thinks, "I have the unique chance, only this week, to push through in parliament a political programme for subsidising carbon-friendly forestry and I cannot wait for future scientific results."

"Nice to meet you and thanks for sharing a beer", both say. But the scientist goes away convinced that the friend has to deepen his understanding of science in order to apply the scientific message to politics. The thoughts of the politician are similar but inverted: The scientist should learn to think a little more like a politician in order to conduct research that is relevant for practice.

It is an often-suggested strategy that scientists and politicians should adapt to each other a little to promote joint solutions. We will investigate whether this is a promising strategy and whether it is possible at all. Maybe we need to accept the general differences between science and politics and learn how to make these differences benefit science-based policy instead.

1.3 Focus on Individual Strengths

The ideal roles of scientists and of political actors are quite clear. Science and politics are public domain tasks and the ideals of both professions have been discussed in public frequently since their very beginnings.

Scientists see their core mission in producing knowledge that is better than other types of knowledge in respect of several important aspects. Scientific knowledge uncovers secrets of nature, enabling us to explain natural phenomena and to intervene in natural processes. For example, the recognition of potato cultivation as having the potential for valuable nutrient production in Europe, and the description of the appropriate soil to do this and of the plant itself, led to changes in agriculture in Europe in the seventeenth and eighteenth century (McNeill 1999). Scientists make a commitment to apply specific scientific methodology to produce such knowledge effectively and efficiently. The enormous output of scientific knowledge has proven convincingly that science serves its purpose and that political actors can rely on a growing wealth of scientific knowledge about the world and of innovative solutions to its problems.

Political actors pursue a different mission. They act in order to solve problems by using political means. They have created specific political organisations like parliaments, public administrations and courts in order to tackle issues using professional and democratic political procedures. The ideal is to organise a public discourse for agenda setting, apply a democratic procedure in formulating the policies and then to implement them via a legally bound public administration. The whole policy should be evaluated by critical media and the democratic participation of citizens, and it should withstand any challenges before the public courts. Political actors produce decisions and formulate and implement political instruments with the aim of contributing to the solution of issues and problems. Political actors at least claim their ideal-typical role to be solving real world problems.

Scientists' ideal role in knowledge production and politicians' counterpart in problem-solving activity, as described above, are very different in reality, though they are linked and they could support each other strongly. Table 1.1 illustrates the differences between these ideals, for both scientists and political actors.

1.3.1 Different Focus of Scientists and Political Actors

Scientists describe and explain the world, whereas political actors change the world. In order to describe nature, one has to suppose that there is something out in the world that has its own reality that can be observed scientifically. Foremostly, one

Table 1.1 Ideal roles of scientists and political actors

Scientists Produce knowledge scientifically	Political actors Solve problems politically
Focus on describing and explaining the world	Focus on influencing and changing the world
Ask critical questions	Ask relevant questions
Collect all data	Select supportive data
Take time needed for analysis	Take time limited by a window of opportunity
Give reversible answers	Give convincing answers
Accumulate scientific knowledge	Label the knowledge and the world anew
Avoid wishful thinking	Mobilise wishful thinking
Rely on theoretical logic and empirical proof	Rely on persuasion and agreement
Use power to describe and explain the world	Use power to change the world

has to observe in the right manner in order to uncover nature's secrets. Scientists do their job by not disturbing the object of their observation, and by making and recording their observations as precisely as possible. From experience they know that good description requires clear terminology. Reliable and meaningful data can only be produced by knowing what to observe and how to observe it, and by writing down all important observations in a protocol.

Even a purely scientific description of a new phenomenon is highly relevant for political issues. To observe and describe a change in climate is a challenging task and a prerequisite for arriving at the right solution. Scientists have to describe the change in temperature: is it the change in the mean temperature, or of the distribution of temperature over time, or of the maxima and minima? Which region is relevant: local forest stands, regional landscapes, nations, continents? Each focus and defined term will have different and complex consequences for the description and needs scientific consideration.

The goal of explaining climate change opens up a wide area of research. Which factors should be analysed—human factors or natural factors? Which theory should be applied—pure climatology or a combination with biology and maybe even social sciences? Which factors can be observed and which innovative methods are needed to observe them? Where does one gather the resources needed for complex research? Scientists who want to propose descriptions and explanations are confronted with a huge set of challenges. They are fully occupied by solving these scientific questions and seek to produce theoretically sound and empirically demonstrable results.

For political actors, observing the world is also helpful, but this does not constitute the essence of their political tasks. Shaping policy means to make a difference in the real world. Political actors claim to find solutions and change the world for the better. They have to *act*. They try to influence the behaviour of society in general, as well as that of other, specific, political actors. Therefore, political actors observe other actors closely. If political actors are able to alter the thinking, decisions and behaviour of other actors, they will be successful in changing the world.

Political actors devote their attention to the whole extent of the policy process. They try to assert their interests through argumentation as agendas are being set. In the case of climate change, they may look for and form "advocacy coalitions" (Sabatier and Jenkins-Smith 1993) to stress the importance of climate change issues. A key factor in this process is to draw the attention of the media. This means that arguments have to be presented in an interesting manner, which causes a preference for shocking news and perpetually new angles, a situation in which the complexity of scientific discourse plays a minor role. In formulating climate policy, political actors have to organise support for specific goals and instruments. They form coalitions and make efficient use of their power sources. The whole process of politics is highly complex and volatile. Successful engagement in politics requires more daily activity than a political actor can achieve in practice. The consequence is that nearly no attention and time is left for seeking information based on scientific knowledge. Henry Kissinger, the famous political scientist and Secretary of State described the deficit of political actors clearly when he stated in his memoirs (Kissinger 1982) that, during his tenure as Secretary of State, he had had no chance to learn anything new in science and that he relied fully on the stage of scientific knowledge he had reached as a professor when he started the new political job. For Kissinger "the qualities that distinguish a great statesman are prescience and courage, not analytical intelligence" (Kissinger 1982, pp. 168–169).

1.3.2 Asking Different Questions

The idea of science is that, through critical research, scientific descriptions and explanations draw an improving picture of reality. Progress in knowledge is achieved by taking nothing for granted and by asking critical questions repeatedly. No scientific finding is excluded from critical analysis in the future, which might well disprove the original results.

For example, a major result in climate change policy was achieved when scientists identified the deleterious consequences of increasing temperature on forests. Research identified threats for forests in southern areas and predicted a serious decline in their stability. The results were alarming, but scientists did not stop with these results. They investigated further so as to find which factors promoted adaptation to climatic changes, and found that among the huge number of trees growing within a specific forest, there were some species and individuals with a higher genetic adaptability to climate change. Therefore, the statements concerning the level of threat to forest health, which were general in hindsight, are more scientifically differentiated today. To predict the consequences of climate change for forests correctly, any scientific evaluation has to consider species and individual genotypes, in addition to other factors.

Critical questions lead to increasingly specific and complex results. In fact, such results again provide a basis for more critical questions. These questions are welcome by scientists, as they are not only interesting for the individual researcher, but also drive scientific progress in general.

Unfortunately, such critical questions regarding existing results are not helpful for political actors. Their task is to build an informational basis that supports specific actions. To reach a consensus on any problem and its preferred solution requires a focus on argumentation. Such focus is fostered by specific questions. For example, in order to mobilise argumentative support to fight emissions, which are accelerating climate change, it is necessary to ask what the specific harmful emissions are and to ask critical questions that can identify their negative effects. But questions that draw attention to other factors causing climate change, or criticise existing judgments on emissions, weaken the consensual basis for political action. Therefore, political actors love questions, but only selective questions that are likely to deliver supportive arguments for them.

In addition, political actors have a need for strong arguments that they can use within the time limit of a window of opportunity (Kingdon 2003). Critical questions that only future research can answer weaken the ability to make a strong argument and to make a decision within a given short time.

The selection of relevant questions to support specific political positions is not due to the habit of political actors. The reason why relevance takes precedence over other factors is that politicians are tasked with making decisions and achieving results. The political decision process is characterised by a narrowing down of the discourse, step by step, to arrive at a story line that supports specific arguments and results in the preferred decision. The wrong questions are a serious threat for this highly challenging process. Therefore, political actors love questions, but only those relevant to the building of support for their chosen position. To be successful, they cannot be open to all critical questions in the way scientists are.

1.3.3 Collecting Different Data

A basis in strong data is essential for good science. Scientists have to be open to all data produced by empirical methods. Unexpected data or data that contradict the hypothesis are welcome. They enable scientists to identify wrong descriptions or explanations and to look for improved ones. Ignoring such data would mean destroying the potential for science to discriminate between correct and incorrect ideas. While it may be disappointing for a scientist to generate data contradicting his expectations, suppressing or ignoring such data is not tolerated in scientific disciplines. The scientific community has evaluation procedures in place to detect possible failures in respect of this type of impartiality in the process of data processing, and to correct them.

Again, the task of political actors is a different one. To move things forward, they have to achieve a consensus as the basis for accepted data. In the political process, such a basis is feeble and always in danger of collapse. Supportive data strengthen the basis and must be gathered exhaustively, whereas contradicting data are harmful and have to be excluded from the political discourse. The most powerful strategy is to ignore such data all together. That is why political actors draw the limited attention of the political discourse strongly to supportive data only. As a consequence, critical data never enters the discourse. For example, in a German case, political actors supported a new scientific expert panel that examined the effects of climate change on forests (Jansen et al. 2008). But the actors of the forest sector lobbied strongly to focus the model on timber production only, neglecting the effects on biodiversity. The lobbyists hoped that if the expert's system showed the effects on timber production clearly, without mentioning effects on biodiversity, the scientific data would support forest policy strongly without having to face critique from nature conservationists (Böcher and Krott 2014b; Storch et al. 2012). The lack of scientific data with regard to nature conservation would have weakened nature conservation interests substantively in the political process. The contradiction regarding the collection of data is similar to the contradiction regarding the questions asked. Scientists need to be open to all—and especially new—data. Otherwise, they would destroy their ability to check their hypotheses empirically. Political actors win by focussing on collecting supportive data. They would risk destroying the achievement of results via consensus if they showed full openness towards new data.

1.3.4 Different Time Frames

Scientists and political actors live within different time frames. Science is an open process looking to the future as a great source for achieving more knowledge. Science considers the time frame for producing new results and improving knowledge to be more or less endless. Science simply *has* all the time in the world. Of course, getting scientific results from a specific project is limited to the schedule of the research project, and milestones should guarantee an efficient progress. But taken seriously, science cannot be planned like a management project. Unexpected results and obstacles are the very content of innovative science. Research projects can even fail because the production of innovative research results cannot be completely planned in advance. Some results might be achieved earlier than planned, some later and some never. In addition, unplanned but relevant results are achieved in many cases. They contribute to the progress of science even if they are not part of the project's rationale. All this contributes to the basic idea in science that time is needed for research, but one cannot predict exactly how *much* and it makes no sense, in light of scientific progress, to stop a research project because a time limit has not been met.

In politics, time frames have another meaning. Issues have to be solved now, and not far in the future. People and political actors live *now* and politics promise to make their lives better. During short election periods (usually 4 or 5 years), other political actors may rise to power and replace the current candidates. Therefore, political actors concentrate on immediate actions, the next election, and tend to neglect the future or the past. Windows of opportunity pop up, sometimes unpredictably, within the political process. Within their time frame, successful action is possible for specific actors. Once they are gone—and this will be sooner rather than later—politics become stagnant again without much chance of moving forward. Political actors are driven by the desire to make the best use of their opportunities and they need scientific knowledge either immediately or never.

As soon as floods are destroying houses and threatening towns, a political discourse appears which has the potential to grow substantially with media support. A political window of opportunity opens, which can be used to formulate new policies and programmes for the conservation of wetlands or the restructuration of settlement plans. Scientific knowledge that is available immediately is welcome and used in political discourse and decision-making. However, no political actor can wait until scientific projects investigate the floods' causes, consequences and suitable instruments to solve similar problems in the future. Political actors can act within a short time window only, and scientists need at least a few years to produce reasonable results.

1.3.5 Revertible Versus Convincing Answers

Science draws a picture of the world that is always limited. But the advantage of this being a picture is that it can be recanted without harming the world. Experience allows science to improve on the picture over time. All scientific answers are revertible in principle. To a great degree, the reversibility contributes to the progress science can make.

If a scientific argument turns out to be wrong, science can reformulate the argument and make progress. If the prediction that a specific kind of genetic breed will protect plants against insects proves false, this will have an impact in practice: plants will die, crops will be destroyed by insects, farmers will go bankrupt and prices for food will increase. Many people will suffer. The consequences are a big mess in practice, but science will not be affected directly. In contrast, scientists will learn from their failures and improve their knowledge, making progress in science.

However, the full burden of the wrong and recanted answers to scientific questions will be borne in practice. Therefore, the most important task of political actors is to convince people that the answers they give are right. The political actors, by linking knowledge to action, change the world. After the action is completed, the procedure is partly reversed at best. In most of these cases, heavy costs are incurred. Therefore, achieving results in politics means persuading people that the proposed solutions are the right ones and that they will work properly as they are. The political actors can overcome the inertia of politics and any preference for staying within established traditions only by demonstrating the strength of the new solutions. They have to argue that the new solution will work properly.

In addition, the implementation of a solution demands strong efforts not only from political actors, but also from society and the economy at large. All these investments will be accepted only if many and strong allies are convinced of the solution. The political actors have to mobilise support for change by convincing people of the necessity to act immediately. Admitting that a particular line of scientific argumentation might be abandoned any time and or that it might be questioned in the future is not helpful.

1.3.6 Accumulating Knowledge or Labelling Knowledge Anew

Progress is based very much on the ability of science to accumulate knowledge. The thoughts and experiences of former generations of scientists or scientists from different countries are accumulated by starting scientific investigations in consideration of the "state of the art". Only when a scientific problem is discussed whilst taking the latest developments into account can new hypotheses be formulated. Close examination of the most recent findings in science is important in order to avoid having to make the same discoveries again and, more importantly, to avoid making the same mistake twice. The limited resources earmarked for research have to be invested in new questions and not in problems that have already been solved. Therefore, in many cases scientific advice will rely to a great extent on existing knowledge and will remind political actors that not every problem in society and nature is completely new. The established body of scientific knowledge is a far better basis for decision-making than new, untested ideas—as promising as they might seem.

In contrast, political actors need new innovative ideas in order to mobilise people for political reforms. Since people have experienced the failure of present policy, presenting alternatives as being "new" is essential to introducing reforms as part of an agenda. Political actors have to highlight the new aspects of problems *and* solutions. New threats grasp the attention of the media and public. Only new solutions can promise to serve all needs, whereas the old solutions are already known for their failures and costs.

If the harvest of an agricultural crop falls behind expectations in one year, scientists will first look to the existing body of knowledge about the factors influencing crop growth. They might find explanations pointing at the weather or harm caused by insects as the culprit for this development. Such "normal and known" causes are not a suitable catalyst for the introduction of a problem into the political agenda. Political actors need to identify either a new dimension to damages that has never been experienced before or a new threat, with unknown negative potential consequences for the future, to generate political discourse around the problem. At present, the projection of local problems onto the global dimension is a technique often used in political discourse to demonstrate the enormous potential threats caused by a specific problem. For example, a suspected genetic mutation in an insect that threatens to spread globally immediately turns a local bad crop situation into a huge problem that underscores the need for political action. Political actors wanting to take action have to "invent" a new world in which the seemingly catastrophic threats can be countered by a new solution. To simply fight the old potato bug populations that exploded in a given year due to favourable weather conditions, for example, would not suffice to mobilise the body politic to favour ecological agricultural reform.

1.3.7 The Power of Wishful Thinking

Wishful thinking is a strong power in politics, but a threat for science. One of the primary objectives of science is to distinguish theoretically and empirically proven arguments from wishful ones. Of course, a scientist may have strong wishes, but his scientific findings should be independent from these. Scientific procedures aim to produce theoretical and empirical arguments that are "intersubjective" in the sense that all scientists would draw the same conclusion given the same hypothesis and empirical data. In this way, the individual preferences of the scientists do not matter.

Politics, however, is quite the opposite of value-free decision-making. It seeks to mobilise people to support specific values and goals. The very driver of politics is the hope and wishful thinking of people and political actors. Taking people's wishes seriously and promising that there are better alternatives in life is a key narrative of politics. Without a good story about how to improve their lives, people will not support policy and policy change. Such a story will inevitably be built upon wishful thinking to a great degree.

At this point, uncertainty becomes relevant for science and politics. Uncertainty is very common in science and part of most scientific arguments. Science has developed the mighty instrument of statistics to describe uncertainty. As a consequence, the difference of scientific arguments compared to other types of knowledge is not the absence of uncertainty in itself, but the ability of science to define uncertainty and to quantify it by statistic estimation.

From the political point of view, the estimation of uncertainty is not very helpful. Political actors have to act—and they very often have only one shot at it. Under this restriction, if the risk of failure of a certain political action is 30 %, political actors cannot act upon the positive 70 % and avoid the negative 30 %. They can only hope to stay within the positive range.

Political actors are rewarded greatly if they can handle risky problems well and make decisions that turn out to have been the right ones in future hindsight. Such a strategy is wishful thinking rewarded by luck and is not espoused by science. The unique historical success is not judged as impossible by science but does not change the probability of failure of 99 %. Therefore, a specific success history will not change the scientific recommendation to avoid such a highly risky solution.

Due to the fact that many political issues and solutions can be described scientifically only with a high degree of uncertainty, political actors can have ample leeway for wishful thinking and can come up with solutions that are only vaguely based on science.

1.3.8 Difference Between Logic and Agreement

A scientific argument has to be based on theoretical logic and on empirical evidence. Everybody who makes a superior theoretical argument or who presents additional empirical evidence can challenge and change existing scientific conclusions by pure argumentation. The majority of scientists agree on the threat of climate change to winter sports in alpine areas. Higher temperatures in winter would reduce the snowfall in alpine regions, resulting in fewer options for good skiing conditions, causing serious economic consequences for tourism. In one of the case studies in this book (see Sect. 3.3) a modelling project of a specific alpine region comes up with different results: the microclimate in this valley will not be influenced much by the overall warming and the snow conditions in winter will stay stable for the next 30 years. As the result is theoretically and empirically sound, scientists have to accept this result and must recant the hypothesis that predicts disastrous consequences of global warming for alpine tourism as a whole. The consequences have been proven to be different depending on the microclimate of each specific valley.

The political point of view is different. Having reached the agreement that global warming threatens the tourism industry is a major political achievement for all political actors who want to propose climate policy. In the country in question, the tourism industry is the greatest contributor to the Gross domestic product (GDP). Therefore, the threat to this key industry is a major argument in the political discourse. The political actors who fight for climate protection and emissions reduction are not willing to weaken their argument of endangered tourism by acknowledging new scientific results. Additionally, their chances are good in that the new scientific results will probably not become strong enough a presence in political discourse so as to challenge the consensus on the disastrous impact of climate change.

In politics, the general agreement on results is more valuable than scientific logic and empirical proof. General agreement supports the position of political actors directly, whereas scientific logic has to be turned into an agreement first to become supportive for political actors. Therefore, political actors orient themselves mainly towards the accepted and agreed-upon arguments. They focus on the belief systems of other actors and the public and try to convince people by demonstrating that a given political argument is well in line with their beliefs. Linking an argument to the beliefs of people and to other, already-accepted arguments, is more important than counting with the backup of scientific logic. Without the possibility of persuading at least a significant portion of the population, scientific logic is useless for political actors.

1.3.9 Power for Describing and for Changing the World

The last point on the political importance of agreement is to do with the most important difference between scientists and political actors. Both groups use power, but in a different way. Scientists use power to describe and explain the world. The main power sources in a democratic political system are the legally-based freedom of research, the well-educated people within the research organisations, these organisations' financial and technical resources and the theoretical and empirical body of knowledge stored within the research organisations worldwide. All these power resources enable scientists to do their job freely, even if the political actors do not agree with their findings. In addition, these power sources enable scientists to contribute to the public discourse and to inform specific actors about scientific results. The scientific ideal states that scientists should use their sources for science only, and not for politicking for specific positions. It would be an abuse of science if a scientist were to make use of scientific results only to support a specific point of view, whilst suppressing critical comments or other findings that challenge his results. In the long run, such a use of power would diminish science's ability to increase knowledge and in the end, ruin science altogether.

Political actors make use of power in a completely different way. Their main task is to change the world for the better and, simultaneously, to improve their own position. Different power sources enable political actors to influence other actors regardless of the latter's will. Opponents can be forced to follow a specific solution. The force of power exceeds that of acceptance by far. Of course, political actors prefer acceptance, but if it cannot be achieved, the political system is willing and able to use power to undermine conflicts, in order to solve problems.

The power sources of political actors are numerous. Political actors can make threats or mobilise force, attract collaborators by offering incentives, or manipulate people using selective or wrong information. Their aim is to change the behaviour of other political actors, and of producers as well as of consumers. The change of behaviour is their proof of success. Scientific arguments play a very small role in this struggle for power. The idealistic hope that scientific truth can influence power ("speaking truth to power", Price 1981; Wildavsky 1979) and have an impact is seldom fulfilled. This is because the very essence of the power of political actors is that they can act and influence solutions regardless of the will of others—including scientists.

A more realistic picture is that scientists speak their scientific truth within a power process. As soon as scientists share their results, they become part of the political process. From that moment on, they can achieve progress only if they are successful within the restrictions and chances of politics.

1.4 Scientific and Political Principles for Bridging the Difference

A close look at scientists and political actors makes visible the serious differences between their respective tasks and roles. It becomes evident that each role has its merits and specific strengths. Scientists are good at describing and explaining the world. They ask critical questions, collect all data and invest the time that is needed for research. The answers of science are reversible and science accumulates its knowledge. Scientists have to avoid wishful thinking and need to rely on theoretical logic and empirical proof. Scientists' power sources have to be used to describe and explain the world more precisely every day.

The focus and strengths of political actors are quite the opposite. Instead of describing it, they are changing the world. In doing so, they ask only those questions relevant to them and they focus merely on supportive data. They need information within the short times allowed by windows of opportunity. Political actors try to persuade people by promising innovative solutions, and making their arguments in consideration of people's wishes and beliefs. In politics, persuasion is more important than logical proof. Finally, political actors make use of power to change the behaviour of other people, including scientists. Therefore, they have a tendency to influence scientists to get their support, rather than to learn from scientists' results.

Decades of studying knowledge transfer tell us that scientists and political actors can meet many times and communicate seriously without diminishing their difference in focus and without adapting their different roles to each other. They are quite stubborn in doing things their way, either as scientists or as political actors. For example, one case in our book (see Sect. 3.6) shows that even if political actors listen to scientists demonstrate the severity of multiple threats for an entire river ecosystem, in the end the political actors conclude that rivers are important for transportation and energy and that flooding has to be regulated in order to provide safe housing areas. This infrastructure element, providing shelter for many people, could not be questioned by ecological fears. This case demonstrates how political actors do not challenge the dominant discourse, nor do they change their favoured solutions due to critical scientific truths.

The authors do not believe that the serious differences between scientists and political actors that block knowledge transfer are caused primarily by the personality of a specific person. They are in fact caused by the different roles the individuals play within the systems of science and politics. Therefore, whilst personal engagement, openness and willingness to learn from each other are helpful, they will never be sufficient to bridge the wide gap between the two systems.

A solid bridge between the two has to admit to these differences, which cannot be diminished without harming either science or policy. If both moved closer to one another and became more similar, they would lose their individual strength. Science would be weaker in observing and explaining the world and policy would lose the potential to regulate conflicts and introduce solutions that could change the world.

By softening the differences between science and policy, we would end up in a weaker system, which would not be based in science and would not be strong enough to make a difference in politics.

Many scientists who became personally engaged in politics were frustrated and describe the futility in attempting to fulfil both roles in an adequate manner. After an initial phase of engaging enthusiastically in a very current political issue closely related to their research area, they experience the loss of the ability to perform good science, as well as the failure to become a successful political lobbyist who dominates the power play. The differences between both tasks are too big to bundle the two into one.

In the last decade, the authors have investigated the practice of knowledge transfer from science to politics in many case studies ranging from the famous German "Federal Environment Agency" (UBA) (Böcher and Krott 2010; Böcher 2010, 2012), the German "Federal Agency for Nature Conservation" (BfN) (Heim and Böcher 2016), state forest research institutes in Eastern Europe (Stevanov et al. 2013), Austrian research programmes for sustainability (Krott 1994, 2002, 2005; Böcher and Krott 2012, 2014a), and state forest research institutes in Japan and Sweden (Nagasaka et al. 2016) to development projects in watershed management in Indonesia (Dharmawan et al. 2016) and community forestry projects in Africa, South America and Asia (Maryudi et al. 2012). Despite the high diversity of the cases, the basic problems for the transfer of scientific results into practice were found everywhere. During the last years, we have designed a model that enables us to describe the specific problems that we found, to evaluate them and to optimise knowledge transfer in the future. This model, called the Research-Integration-Utilization (RIU) model, provides the theoretical basis for this book.

The principles of our RIU model are based on the acceptance of the fundamental differences between science and policy. The model looks for a partial integration that keeps these differences, whilst building on them in order to offer a bridge for the transfer of knowledge. The principles could guide scientists and political actors if they are interested in improving the transfer of knowledge.

The RIU model defines four principles to achieve integration of science and policy:

RIU 1 Building trust in encounters
RIU 2 Exchanging a maximum of information
RIU 3 Accepting limits of mutual understanding
RIU 4 Looking for allies of science

1.4.1 RIU 1: Building Trust in Encounters

Scientists and political actors must be able to meet in order to exchange information. Creating new opportunities to do this is a requisite in many knowledge transfer models. For example, Gibbons et al. (1994) created a "Mode-2" model of knowledge production that would include new types of encounters between science and the public, which they called "*agora*", after the term for the assembly areas of ancient Greece (Nowotny et al. 2001, p. 201 ff.). Like in Greek antiquity, scientists and political actors meet regularly to debate their issues completely freely and trustfully. Such meetings are only productive if both sides trust each other. Scientists and political actors must accept each other as partners.

Building up trust between such different systems as science and politics is not an easy task and has to overcome many conflicts. However, increasing and reaching an optimal level of trust is necessary to exchange relevant information. Therefore, building up trust requires professional management.

It is helpful for scientists and political actors to get to know each other personally. Tolerance and intercultural experiences are required to avoid conflicts caused by misunderstandings. For example, scientists' habit of discussing topics critically on one side and the affirmative language of political stakeholders on the other make it difficult for them to talk about the same thing, even if both basically agree on a certain topic.

Joint efforts in solving a problem increase the level of trust. Therefore, a common background in a political sector, or joint political engagement, is helpful for building up trust. Most sectors, like agriculture or environmental protection, have their own research organisations and corresponding departments within universities that are devoted to the same mission of finding solutions for agriculture or the environment. Such organisational frameworks create a good basis for trustful encounters of scientists and political actors.

Finally, the personal behaviour of scientists and political actors has a big influence on trust. In general, a basic level of trust between scientists, political actors and the public already exists in democratic societies. However, it can be easily destroyed by imprudent behaviour. As an example, political actors who use scientists to lobby for their causes will disappoint them for a long time. Scientists who let themselves be driven by the public discourse into making value-laden statements that are not scientifically sound will lose the trust of the scientific community and the public soon. Therefore, in pluralistic democratic societies, trust is not something scientists or political actors have to gain, but something that should not be lost due to individual behaviour.

1.4.2 RIU 2: Exchanging a Maximum of Information

Even if both sides trust each other and are strongly interested in good communication, the fruitful exchange of information between science and policy is still not guaranteed. A big language problem separates science from politics. Science and politics have developed their own languages that differ strongly. Scientists love clearly defined terms and complex arguments, rich in technical details. The political language consists of vague, but value-laden terms and simple messages that avoid technical details and links to other arguments, in order to be easily understood.

To learn and practice the language of the other side is necessary for the flow of information. Scientists must develop a familiarity with the political terms and learn how to cover political issues using scientific terms. The term “biodiversity” is a big catchword in environmental politics, signalling the needs and the engagement of modern nature conservation. However, establishing a link to its scientific meaning is rather hard. Is biodiversity the number of species, the scarcity of a specific kind of species, the diversity of ecosystems, the genetic variety, or all of these together? What is to do if the different aspects contradict each other and, for example, increasing the variety of species and the protection of one specific species need opposite strategies? The precise scientific language is able to figure out and communicate all the aspects of the term “biodiversity”, whereas the political language will never be more precise, and will simply rely on the general term “biodiversity”. Instead of defining a precise term, political actors will link biodiversity to value-laden terms like “our national” or “global heritage”, or “endangered”.

Both partners have to adapt their language somewhat in order to provide some orientation for the other side. A major risk for misunderstandings is that scientists often expect clear and meaningful goals to be formulated by political actors. There are a lot of goals being mentioned in political discourse, but they are all formulated vaguely and are value-laden. Political goals have a symbolic side that communicates trust, which is just as important for politics as a rational orientation (Edelman 1967). Meaningful information can only be exchanged between political actors and scientists if scientists accept this symbolic side of goals and do not expect scientific clarity and priorities from political actors. Nevertheless, scientists can still recognise some content in the issues and preferred solutions within the setting of the vague political goals.

On the other side, the political actors must realise that scientific knowledge is either clear or does not exist at all. A scientific statement about the vague term “biodiversity” is not possible. Scientists have to define the term “biodiversity” first and then the scientific statement is relevant for this definition only.

A fully open exchange of arguments is the best way to increase the flow of information. Additionally, both sides have to take sufficient time to listen to arguments that are not fully comprehensible to them. The whole story should be told, as one side might find some details to be very interesting that another side would classify as irrelevant. For example, on the topic of “forest dieback”, a scientist might mention, among other negative factors caused by climatic conditions and emissions, a nutrition deficit. This detail will immediately catch the attention of a political actor that is close to the emitting industries. The proposed solution to subsidising the fertilisation of the forest publicly is greatly admired by a political lobbyist. This message might be the only one making the transition from science into the political process—but this is better than nothing.

Even if far from all of the information were to flow between science and policy, learning the other side’s language, taking time to listen to every argument—even if some are incomprehensible—and accepting a selective flow of information are principles that improve the information exchange as much as possible.

1.4.3 RIU 3: Accepting Limits in Mutual Understanding

If we recognise that even the information exchange between science and political actors is very limited and selective, how we can expect a transfer of knowledge at all? The answer is that one does not need to understand the partner *fully* in order to get some very valuable information. The only requirement is that the information from one side can be correctly linked to the information needs of the other side.

Figure 1.1 shows the basic communication lines between scientists and political actors.

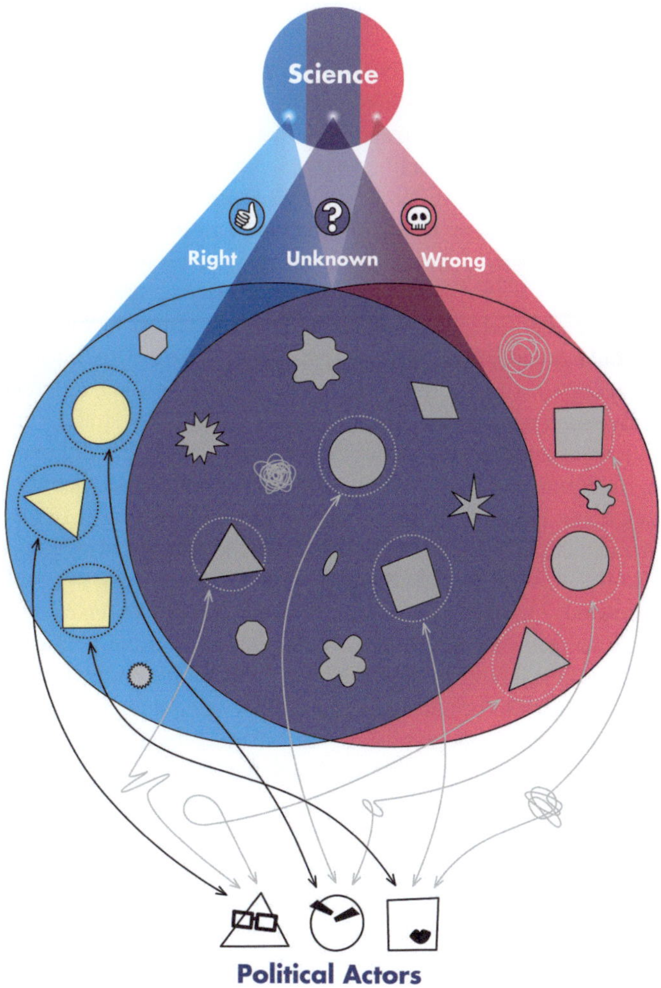

Fig. 1.1 Selection of scientific knowledge

With regard to a specific problem, the scientist formulates some results, their development, causes, and options for solutions. The scientist's key task is to discriminate between correct and incorrect statements ("right"/"wrong" in Fig. 1.1). As well, the scientist identifies statements about problems to which scientific solutions are currently uncertain or not available ("unknown" in Fig. 1.1). The political actor is neither willing nor able to follow the scientific reasoning used to distinguish correct and incorrect ideas. He also does not care if scientific solutions are unknown. Instead, his competence lies with the ability to recognise the link of the statements about the development of the problem, its causes and possible solutions, to his own interests and public goals. Immediately upon detecting interesting results, he will listen and get the gist of the basic message. He cannot check for the scientific strength of the results, but he can select those parts of the message that are useful for him (represented by the three different geometrical forms in Fig. 1.1).

An example is mentioned in this book (see Sect. 3.6): A scientist had conducted a lot of research on the ecology of different types of endangered rivers. He communicated the results to the ministry responsible for the protection of rivers. Although the ministry had financed the scientific analysis, it was not very interested or willing to consider any political actions that would cause serious conflicts with the users of these rivers. Years later, the protection of water systems in European Union member countries was set as part of the political agenda and each member country was obliged to develop a concept for the protection of its water systems. Immediately, the attention of the ministry changed and the bureaucrats started looking for scientific results that would support their political task. Under these new political conditions, the scientist received a lot of attention for her results and her different types of endangered rivers even became part of the country's new water strategy and law. It is important to mention that the scientific message about the different types of endangered rivers remained the same over the whole period. The scientific reasoning did not change, but the interests and needs of the political actor—in this case, the ministry—shifted and drew the attention towards the ecology of rivers.

This example illustrates that scientific reasoning alone is not sufficient for achieving knowledge transfer. The scientific *results* have to be *linked* to *interests* of a specific political actor in order to be able to expand into the political discourse. The link to the political actor is the issue, not the scientific content of the research results. Both sides need to be able to recognise the common issue, even if they see it from a different angle at the beginning. In our example, the issue is the protection of rivers and the scientist defined and analysed the river based on the theory of ecosystems. The ministry did not get into the ecological theory, but recognised that the scientific results were highly relevant for the design of the political protection strategy for rivers, and consequently used the scientific results.

The political actor is not able to understand the scientific arguments and make his own scientific judgment. He has to rely on the scientist and trust his message. This means that the scientist is fully responsible for the scientific results. He cannot share the responsibility with the political actor by telling him the scientific arguments, theories and empirical data. The scientific quality can only be provided by the scientist who follows scientific standards and who is part of the evaluation procedures of the scientific system. The political actor is neither willing to become part of the scientific system, nor does he have the ability to do so. Therefore, he lacks the competence for checking the correctness of a scientific result. The decisions regarding correctness are made solely on the basis of scientific procedures.

The absence of scientific competence does not mean that political actors cannot draw conclusions from a scientific statement. They can very well recognise its political potential. Political actors are highly competent in linking information to the political discourse to support their interests. If the scientific statement is clear and shown to be correct, the political actor will select it easily as being either helpful or unhelpful for his political agenda. The political actors cannot choose statements that have been disproved scientifically. Statements which are scientifically right, but do not seem useful from the political actors' point of view are ignored as well. Science has no means to push such statements into the political discourse against the will of political actors.

But what happens if the scientific results are not clear and science is not able to diminish the uncertainty? Figure 1.1 shows the options for knowledge transfer in such a case. Scientists then have to inform about the results and, in many cases, they are able to describe the uncertainty statistically. That is all science can do. As mentioned before, political actors have to take *action*, which means that the uncertainty has to be transformed into an action that tries to solve a problem by changing the world. Once the scientific information includes uncertainty, political actors have to make their own judgment, turning the uncertainty into an action or not. Scientific uncertainty increases the responsibility of the political actors for their actions. Scientists are unable to take over this responsibility, because science can only deliver what is described and measured by uncertainties.

A lack of competence to understand each other is a fundamental obstacle that makes a complete knowledge transfer almost impossible. Progress can be achieved only if the transfer process enables both sides to handle the knowledge deficits and to organise a productive exchange despite limited mutual understanding. If both sides accept their limits, they can focus on their key competences. Scientists should focus on judging the scientific soundness of the results and political actors should focus on judging the relevance for their interests and public goals. The specific issue provides a link between the different worlds of science and politics.

1.4.4 RIU 4: Looking for Internal and External Allies of Science

The different roles of scientists and political actors open up a huge potential for knowledge transfer exceeding by far the common practices of consensus and of mutual persuasion. This potential can only be developed by scientists who do not try to play a political role themselves.

The simple transfer happens when political actors become aware of scientific problems or solutions and use them to achieve their political goals. The chance for a positive transfer increases if political actors approach scientists in a trustful, mutual process of information exchange. This way, the political actors become internal allies for science. They trust in science to provide a better understanding of the problems and to increase the number of alternative solutions, with a small risk of failure. The scientists in turn are able to look closely at the relevant political issues from the internal ally's position, including preferable solutions and political restrictions for new, scientifically designed, alternatives. Due to the stronger power sources, the political ally in the end decides which parts of the scientific results become parts of a certain policy and which do not, but science ensures that these parts are scientifically sound.

The complex transfer happens when political actors do not engage with science directly, but rather trust the fruitfulness of the transfer of scientific knowledge to other actors. A political actor has the chance to force other political actors to cooperate closely with science. In this case, the actor becomes an external ally of science. He pressures other actors to discuss their problems with scientists, to be open to new results and to adapt their policies to new scientific findings. In this case, a third actor who plays the role of an external ally forces the political actors to pay attention to science.

External allies of science are quite frequent and effective. In our example of water protection (see Sect. 3.6), the European Union was the strong external ally who forced the Austrian National Ministry to take the scientific results about the ecological types of rivers seriously and to consider them as a basis for national water policy.

External allies have a huge potential for fostering knowledge transfer from science to political actors. They do not need any kind of scientific competence themselves, nor do they need to invest in the complex process of contacting with the scientists. They simply force other political actors to do this difficult job. Furthermore, external allies guarantee that political actors take scientific findings seriously. They even force political actors to show whether they have engaged scientific expertise and followed scientific recommendations. If political actors prefer a policy other than the one suggested by science and its external allies, they have to present good scientific arguments for their alternatives, or they will be forced by the external actors to follow the solutions brought forward by science.

Allies provide power to the transfer of knowledge. Two aspects of the relationship between power and science are important: First, power that supports scientific findings does not necessarily contradict the scientific rationale; and second, power can deliver scientific knowledge deep into politics, limited only by the number of power sources.

Political power and scientific rationale seem to contradict each other, because a scientific argument is built purely on logic and empirical evidence. Within the scientific rationale, there is no space for power. But the knowledge transfer reaches beyond the sphere of science into the political system. In politics, power becomes the crucial source of efficacy. Therefore, a political actor who puts his power behind specific scientific arguments in order to force other political actors to base their decisions and acts on these scientific arguments supports the transfer of scientific findings. The fact that the "weaker" political actor is not convinced of the validity of the scientific argument and accepts it only due to the pressure imposed by the stronger actor does not diminish the scientific truth. If science demonstrates the positive effects of using seatbelts in cars, for example, and policy prescribes the use of seatbelts, the lives of both types of drivers are saved: those who believe in them and those who do not believe in the functionality of seatbelts, but are nonetheless forced to use them. The political force does not disturb the scientific truth, but merely fosters its application. Power can transfer scientific results within the political system. Therefore it is not a necessity to base the whole process of scientific knowledge transfer solely on scientific arguments.

Power has a huge potential for transferring scientific knowledge into politics. Both internal and external allies of science put their power behind scientific findings and force other political actors to make use of them. Political allies need not be the most powerful actors in the political system to be successful. It is sufficient if they are stronger than the specific actor they seek to influence. If the owner of an enterprise wants to force her management to use new, scientifically designed methods to improve efficiency, she can do so for her own enterprise even if, compared to bigger enterprises, hers is very small.

Whenever one actor in society, economy or politics is stronger than another, he can use his power dominance to push scientific findings forward and force the weaker actors to act according to the scientific recommendations. Figure 1.2 shows this promotion of scientific findings within society, carried out by means of power, in an unbalanced way and on multiple fronts. The different sizes of the bubbles represent the different strengths of actors. The stars represent scientific knowledge, advanced by more powerful actors. The triangle symbolises knowledge that is not science-based.

The decisive factor for successful knowledge transfer is not the absolute amount of power as such, but the relative dominance of power compared to a specific actor. Every unbalanced power relation opens up the chance for effective transfer of scientific knowledge into actions. The dominant teacher transfers the logic to the student, the dominant employer to the employees, the dominant bureaucrat to the clients, the dominant expert to the laymen, the dominant doctor to the patient, the dominant government to the citizen, the dominant states to the weak states and so on. The world is full of unbalanced power relations and each offers the potential for an effective transfer of scientific knowledge.

To acknowledge that power can support the transfer of scientific knowledge does not mean one neglects the influence scientific knowledge can have when simply shared with political actors. Such processes can add support to the knowledge transfer. However, science would miss a big chance for knowledge transfer if it did not make professional use of the link to power. The informational transfer processes would be closer to the world of science, but their impact in the political sphere would not be the same.

Fig. 1.2 Unbalanced promotion of scientific knowledge for its application in practice, by means of power

Both internal and external allies can become learning allies too. Learning allies reconsider their own beliefs and interests in the light of new scientific findings. If science discovers that spruce trees which promise fast growth and good profits in general will be infected by fungi after the age of 50 due to moist soil and warm climate conditions, why should the forester not revise his original interest in planting spruce, and switch to a reasonable share of beech, which will grow healthily even at old age? There is a potential for accepting scientific findings if an open-minded actor listens to the scientific arguments. We call such an actor a "learning ally" of science, in our RIU model. The notion of such an ally may be an idealisation on behalf of scientists, or it may be an overestimation made in political programmes, but, in principle, such a role is indeed possible in the RIU model. The model provides detailed information as to how such learning allies might be integrated into the knowledge transfer process.

Finally, a "wise ally" is one step further in reconsidering scientific knowledge than the learning ally. He listens carefully to all scientific arguments and compares them with the knowledge gained from other sources, like his own experience or traditional knowledge. The wise ally considers all kinds of knowledge on an equal level. He does not give scientific arguments priority over practical experience. He takes full responsibility for drawing his own conclusions. In the case of a contradiction between scientific arguments and practical experience, he judges autonomously which argument should be accepted.

The RIU model shows exactly the role of the wise ally. The scientific rationale is only *one* alternative for the wise ally. He examines scientific arguments from a broader point of view and comes up with a decision based on a wide range of knowledge. If his conclusion contradicts the scientific arguments, this can still be accepted by science, because the wise ally refers to another type of knowledge. His independent decisions do not question the scientific system. The only argument he cannot use is to claim that his decision is fully based on science.

The wise ally becomes highly relevant within all participatory processes that give the final decision rights to non-experts. They might vote for an alternative solution relying on different types of knowledge and experience and, in addition, take full political responsibility. However, they need to be aware that in those cases, their decisions can no longer be legitimated by science. In practice, most decisions are taken by people who take responsibility for their lives and do not follow all scientific recommendations. They follow mainly the model of "wise allies". Whether they are really "wise" depends on the degree to which they are aware of the scientific arguments and how good their additional sources of knowledge are. The wise ally is convinced that scientific knowledge is only one of many types of knowledge and sometimes it might be better to follow other sources. Since he is an independent actor, science cannot tell the wise ally how much scientific knowledge he should use. He might even completely disregard any scientific knowledge. The RIU model is able to measure how broad the scientific basis for the wise ally's choices is, but it does not claim that he has to meet a specific extent. In this sense, the wise ally is always on the borderline of making unscientific decisions and of showing the limits of science.

1.5 Professional Organisation of Knowledge Transfer

The principles of knowledge transfer need to be organised in order to make a practical contribution. Scientists and political actors can meet in many different settings. The entire process can even happen within a single person if he or she works both as a scientist and as a political consultant. Knowledge transfer can also take place within a small organisational unit that engages in science and consulting. An organisational option that is well developed in practice is found in organisations that are dedicated to "applied research".

Another setting where knowledge transfer takes place is the meetings of scientists and political actors. This can be at a conference, a regular meeting or a transdisciplinary research project that aims to integrate actors and to produce knowledge with practical relevance. Whole research programmes are devoted to inter- and transdisciplinary research for solving big problems, like climate change or sustainable transportation.

For most public issues, specific groups are created and dedicated to the task of knowledge transfer. Such committees might consult the public, the parliament or public administration. In the private sector, enterprises organise huge facilities for research and development of innovative products.

The organisational options are manifold. But all organisations have to fulfil specific tasks in order to transfer scientific knowledge successfully. The RIU model describes those tasks and provides a basis upon which to define standards of quality to meet the challenges of bridging the two very different spheres of science and policy.

1.6 Components of the RIU Model

The RIU model consists of the three components: research, integration and utilization. The production of knowledge on a specific issue has to include all three components in order to be scientifically based and applied in practice.

The first component is scientific research. Here, knowledge is produced by scientific methods following scientific standards. The main tasks are to identify the research question, to define the subject of research, to develop methods, to gather data in the field and to draw theory-based scientific conclusions. All quality standards are defined by the scientific system, and meeting these is a prerequisite for successful knowledge transfer. Having the goal of producing scientific knowledge that can be used for problem solving does not imply in any way a diminution of scientific rigour or quality.

The second component is integration. In this phase, the research is oriented towards practical issues and needs. Integration is a hypothetical process in two directions. Specific issues provide guidance for the formulation of scientific questions and scientific results provide guidance for solutions in practice. Integration is done hypothetically, in the sense that both sides *imagine* what could happen and what could be done through research and in practice. Integration looks at public goals inside the policy process as a well as at the specific interests and power sources of political actors.

The third component is utilization, where scientific results are used to solve problems in practice. Actors within the practical sphere carry out utilization, and science no longer has an influence on what should be done and how the practical sphere is to understand the scientific information. Utilization is carried out by the multiple actors of the private and public spheres. Among them are actors like the parliament, public administration, private enterprises, or citizens. The pluralistic democracy might provide quality standards for this process, but the RIU model also works in other political systems.

1.7 Effect of the RIU Model

The RIU model links the three components for successful knowledge transfer. A specific issue or problem meets the requisite three components. The process might start in the political realm if the public discourse identifies an important issue. Within the integration phase, the issue is examined with regard to public goals, political actors that might be interested in them and the options within the specific policy process. Additionally, the problems will be translated into scientific questions. The existing body of knowledge and the available research methods will be examined. The whole process will result in a good estimate for those scientific problems that have great practical relevance. The process is selective in the sense that not all relevant aspects can be translated into research and that not all relevant arguments can be accepted as being scientifically sound. Integration is also highly innovative in that it questions the existing research questions *as well as* the existing arguments in the field.

Integration delivers selected and more specific problems into the research component. Only those problems that have passed the filter of integration become research subjects. The scientist can now fully engage in research, taking into account all scientific standards. If political actors intervene to shape the results, the scientific quality is diminished. The RIU model stresses that research needs its protected realm.

After a period of research it is useful and necessary that the intermediary results should be discussed within the integration component again. This is where the relevance for the political realm is checked again in advance. Integration can adapt the scientific questions or the methods if the designed research project does not meet the needs of political actors, the policy process or public goals. In addition, new results can catch the attention of political actors that have not been involved yet.

The adapted and newly specified problem enters a research phase again. It has to remain there until some scientific results are produced. The amount of time an issue remains inside the research phase component depends on the research process and the scientific standards alone and it is independent from the wishes of political actors.

After going back and forth between research and integration a few times, some results are ready to be delivered for utilization. The science-based and understandable information is handed to political actors. In the utilization phase, scientists no longer have direct influence. The responsibility for a democratic, legal and effective solution for practical issues switches to the political system.

The only part of utilization in which scientists are directly involved is the introduction of scientific results into the scientific discourse. Here, scientists are responsible for presentation, and they check the results within the scientific community, meeting specific standards of rationality. The RIU model does not expect political actors to follow such scientific standards if they took part in the discourse. Therefore, the space for scientists to encounter political actors is not in the course of scientific discourse, but during integration, where practical discourse and actors might contribute to the hypothetical estimation of an issue's relevance. The empirical test of relevance is the utilization that excludes scientists.

The RIU model serves to show both scientists and political actors their *own* realm of specific competences and responsibilities, as well as the *shared* competences and responsibilities. The RIU model requires from both scientists and political actors to meet actively within the integration component, to accept the selection process undertaken inside the integration component and to renounce interventions inside the research or utilization components. If both sides focus on their specific individual and joint competences and responsibilities, the transfer of knowledge between the spheres of science and policy will be most effective.

The following chapter elaborates on the RIU model of scientific knowledge transfer, which will serve as a guide in our case analyses in the subsequent chapters.

References

Böcher, M. (2010). Federal Environment Agency (Germany). In D. Guston & J. G. Golson (Eds.), *Encyclopedia of Nanoscience and Society* (pp. 234–235). Thousand Oaks: SAGE.

Böcher, M. (2012). Wissenschaftsbasierte Politikberatung auf Abruf – Zur Rolle der Ressortforschungseinrichtungen für Ministerien am Beispiel des BMU. der moderne Staat, *5* (2), 459–480.

Böcher, M., & Krott, M. (2010). Umsetzung des Konzepts einer modernen Ressortforschung im Geschäftsbereich des BMU (Vol. 39/2010, Umweltbundesamt-Texte). Dessau-Roßlau: Umweltbundesamt.

Böcher, M., & Krott, M. (2012). Nachhaltigkeitsforschung in Österreich. *GAIA–Ecological Perspectives for Science and Society, 21*(2), 155–157.

Böcher, M., & Krott, M. (2014a). *Mit Wissen bewegen! Erfolgsfaktoren für Wissenstransfer in den Umweltwissenschaften*. München: Oekom.

Böcher, M., & Krott, M. (2014b). The RIU Model as an Analytical Framework for Scientific Knowledge Transfer—the Case of the "Decision Support System for Sustainable Forest Management within Climate Change". *Biodiversity and Conservation, 23*(14), 3641–3656.

Dharmawan, B., Böcher, M., & Krott, M. (2016). Failure of science-based win-win solution in fishery management: Learnings from Segara Anakan Waters, Central Java, Indonesia. Under review.

Edelmann, M. (1967). *The symbolic uses of politics*. Urbana et al., University of Illinois Press.

Gibbons, M., Limoges, C., Nowotny, H., Schwartzman, S., Scott, P., & Trow, M. (1994). *The new production of knowledge: The dynamics of science and research in contemporary societies*. London [u.a.]: Sage Publications.

Heim, J., & Böcher, M. (2016). CITES and science: Using the RIU-model to analyze institutionalized scientific policy advice in Germany for the case of ivory trade. *Journal of International Wildlife Law & Policy, 19*(2), 159–175.

Jansen, M., Döring, C., Ahrends, B., Bolte, A., Czajkowski, T., Panferov, O., et al. (2008). Anpassungsstrategien für eine nachhaltige Waldbewirtschaftung unter sich wandelnden Klimabedingungen-Entwicklung eines Entscheidungsunterstützungssystems. *Forstarchiv, 12* (4), 131–142.

Kingdon, J.W. (2003). *Agendas, alternatives, and public policies* (2 Edn, Longman classics in political science). New York [u.a.]: Longman.

Kissinger, H. (1982). *Years of Upheaval*. London: Weidenfeld and Nicolson.

Krott, M. (1994). *Management vernetzter Umweltforschung: wissenschaftspolitisches Lehrstück Waldsterben*. Wien: Böhlau.

Krott, M. (2002). Evaluation of transdisciplinary research. *Encyclopedia of life support systems (EOLSS), 3*.

Krott, M. (2005). *Controlling für Programmforschung (Forschungsprogramm Kulturlandschaft)*. Wien: Bundesministerium für Bildung, Wissenschaft und Kultur.

Maryudi, A., Devkota, R. R., Schusser, C., Yufanyi, C., Salla, M., Aurenhammer, H., et al. (2012). Back to basics: Considerations in evaluating the outcomes of community forestry. *Forest Policy And Economics, 14*(1), 1–5. doi:10.1016/j.forpol.2011.07.017.

McNeill, W. H. (1999). How the Potato Changed the World's History. *Social Research, 66*(1), 67–83.

Nagasaka, K., Böcher, M., & Krott, M. (2016). Science-policy interaction: Who does what? Case of the Forest and Forestry Revitalisation Plan in Japan. Under review.

Nowotny, H., Scott, P., & Gibbons, M. (Eds.). (2001). Re-thinking science. *Knowledge and the Public in an Age of Uncertainty*. Cambridge [u.a.]: Polity Press [u.a.].

Price, D. K. (1981). The spectrum from truth to power. In T. J. Kuehn & A. L. Porter (Eds.), *Science, Technology, and National Policy* (pp. 95–131). Ithaca; London: Cornell University Press.

Sabatier, P. A., & Jenkins-Smith, H. C. (1993). *Policy Change and Learning: An Advocacy Coalition Approach (Theoretical Lenses On Public Policy)*. Boulder, Colo.: Westview Press.

Schnitzler, A. (1993). *Professor Bernhardi and Other Plays (Studies in Austrian Literature, Culture, and Thought Translation Series)*. Riverside, CA: Ariadne Press.

Stevanov, M., Böcher, M., Krott, M., Krajter, S., Vuletic, D., & Orlovic, S. (2013). The research, integration and utilization (RIU) model as analytical framework for the professionalization of departmental research organizations: Case studies of publicly funded forest research institutes in Serbia and Croatia. *Forest Policy And Economics, 37*, 20–28. doi:10.1016/j.forpol.2013.03.006.

Storch, S., Krott, M., & Böcher, M. (2012). Institutionalization of accountability within mission-oriented research: The example of the 'Decision Support System Forest and Climate Change'. In K. Hogl, E. Kvarda, R. Nordbeck, & M. Pregernig (Eds.), *Environmental Governance: The Challenge of Legitimacy and Effectiveness* (pp. 260–279).

Wildavsky, A. (1979). Speaking truth to power. *The art and craft of policy analysis* (1 print edn.). Boston [u.a.]: Little, Brown and Co.

Chapter 2
The RIU Model as an Analytical Framework for Scientific Knowledge Transfer

2.1 Introduction to the RIU Model

Like the two friends sharing a beer in the first chapter, many researchers in the environmental sciences experience frustration because practical actors do not follow their scientific advice. Conversely, many politicians are frustrated with the naivety of natural scientists regarding the potential for political implementation of their recipes. The widely-assumed incompatibility between science and politics is the starting point for our analysis: We question whether there may be possibilities for successful scientific knowledge transfer in spite of the fact that the types of logic that political and scientific systems use are different.

In this chapter, we introduce our model of scientific knowledge transfer, which we call RIU, and which stands for research, integration, and utilization. The analysis in the six case studies that follow in Chap. 3 is based on this new RIU model of scientific knowledge transfer.

An understanding of the factors involved in knowledge transfer supports this model. These include state-of-the-art, independent research and the effective, practical utilization to which it leads. In addition, the model highlights the important "integration" activities. Here, the word "integration" refers to the active, bi-directional selection process of those research results that are relevant in practice. The term as a whole emphasizes the importance of special allies in practice who are powerful actors that support the use in society of the results that scientific research achieves.

The RIU model goes a step beyond the existing criteria used to determine the success of knowledge transfer, extending these to include the actors who are the recipients of scientific advice, and their interests and power (Böcher and Krott 2014a, b; Krott 2012). Moreover, we follow recent studies on the use of scientific

M. Böcher and M. Krott, *Science Makes the World Go Round*,
DOI 10.1007/978-3-319-34079-1_2

findings in political practice that speak of a "politicisation of expertise" (e.g. Weingart 1999; Hoppe 1999) and conclude, given the different interests of the actors involved, that it is important to couple the demand of various actors actively with the available supply of expertise (Sarewitz and Pielke 2007; McNie 2007; Guston 2001). The RIU model focuses on three key factors for successful scientific knowledge transfer. By distinguishing between the three elements of research, integration and utilization, the RIU model focuses specifically on the logic of each element. Its main hypothesis is that an integrative selection process based on specific political or practical demands can link the observable and mostly independent action spheres of scientific research and politics (remember the conversation of the two guys having a beer…). This "integrative selection" step is crucial to our model, and we refer to it as integration because a selective integration process is always necessary to connect the worlds of science and politics in order to find science-based solutions for policy and practical problems.

The RIU model relies on the criteria of relevance, credibility and legitimacy (Cash et al. 2002; Mitchell et al. 2004) as applied to specific actors who play a role as allies of knowledge transfer. Here, the connection between scientific knowledge and the concrete demands and interests of the actors is especially important, as is also the link between scientific research results and the interests of actors in the political process (Böcher and Krott 2014a, b; Krott 2012). In addition, there is a need for an intermediation strategy oriented towards target groups (Müller-Rommel 1984; Renn 2003). This strategy should identify the media and formats needed for different addressees (e.g., science, political actors, citizens, administration). Each type of addressee needs its individual kind of science-based information and communication.

What is novel in our model is that it reflects many aspects of scientific knowledge transfer other than mere "communication". Many guidelines for scientific knowledge transfer often state that mostly all that is crucial for knowledge transfer is communication (see for example: Government Office for Science 2010; Moll and Zander 2006, pp. 59–72; Ugolini et al. 2015). Certainly, we agree that science communication in an important subtask of knowledge transfer. However, our political science background leads us to suggest that a much more important aspect to consider is how to connect science-based information to the resources (political power, finances) of political and practical actors. If science-based information is useful to the actors and their interests, then they will use their resources to promote the use of scientific knowledge in their specific realm.

The RIU model consists of three activities

Research,
Integration,
and Utilization.

Research applies scientific principles, methods and standards. For there to be knowledge transfer from science these research results have to be put to the service of the needs and demands of the practical allies and their interests. The requirements

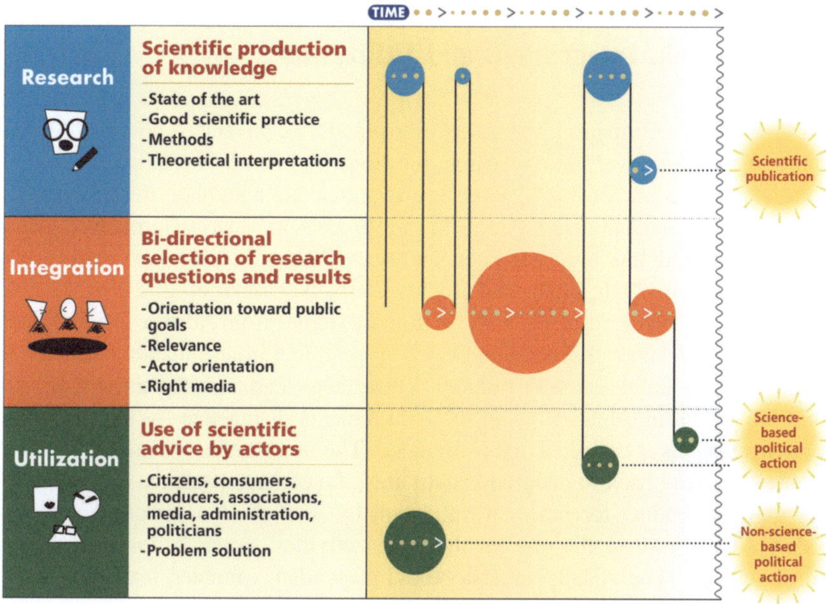

Fig. 2.1 The RIU model of scientific knowledge transfer

of political and practical problems and their solutions determine the criteria for selective integration and practical utilization.

Figure 2.1 shows the RIU model and the three activities necessary for scientific knowledge transfer, namely research, integration, and utilization, each with its respective main criteria for quality. The diagram also shows that scientific knowledge transfer processes take time. During this time period several switchovers between the different activities (from integration to research and vice versa) take place. These involve each of the activities to a different extent as they refer to the specifics of the individual knowledge transfer process and the underlying policy or practical problem. Not every political action is based on scientific knowledge, and our RIU model reflects this, in that it includes a category called "non-science-based political action". Here, political decisions are not based on processes of research and integration. In other cases, continuous research and integration activities lead to science-based political action. Such processes differ in the amounts of research, integration, and utilization involved at any given time. In the figure, the circles represent this in that they have different sizes that correspond to the different lengths of time needed for each activity. The figure illustrates one of the basic ideas of RIU: Science-based political action has to be the result of research and of integration activities, whereas non-science-based political action does not need these. The products of RIU processes can be political and practical actions, by different actors, that use scientific advice. Another product, authored by the scientists involved, can be scientific publications that target the scientific community.

The following sections describe the model, its foundations and its analytical power in detail.

2.2 Scientific Knowledge Transfer and Its Subtasks: Research, Integration, and Utilization

How is scientific information "translated" into practice? Conversely, how are practical requirements "translated" into research questions that can be investigated through scientific research processes? We observed that there are a number of steps between research activities and their integration, in various cases in Austria, Germany, Croatia, Serbia, Japan and Indonesia (Böcher and Krott 2014a, b; Stevanov et al. 2013; Dharmawan et al. 2016; Heim and Böcher 2016; Nagasaka et al. 2016). In later chapters we present the results of some of the case studies from Austria, in which such steps have been taken within certain research programmes in sustainability research or within departmental research institutions. These steps lead from the activity of integration in which information about the requirements of practice for science-based problem solutions is collected and formulated. The next step is to conduct research activities that lead back to integration activities, if a concrete, science-based knowledge transfer product for practical utilization by politicians or practical actors is needed. At the end of this process the three activities, research, integration, and utilization (RIU), analytically understood as independent variables, lead to an active use of science-based information by different political and practical actors.

"Production lines" reflect the time dimension in this process (in Fig. 2.1 these go from left to right along the x-axis) and they reveal a typical alternation between integration and research activities over time, making clear that a science-based product for political and practical utilization must always comprise both research and integration (representing the demands of practice). This is a central prerequisite for successful scientific knowledge transfer. Integration is thereby the bi-directional orientation of scientific knowledge towards the solution of practical problems, and vice versa. Integration is a specific activity that is closely connected to science and that connects science to practice and its requirements (Böcher and Krott 2010, 2014a). Along the timeline of scientific knowledge transfer, there are steps that represent research activities, others that represent integration activities, and steps of a third kind that represent utilization activities.

What is scientific knowledge transfer?

According to our RIU model, we define scientific knowledge transfer as follows:

> Scientific knowledge transfer is the connection between research, *and* integration, *and* utilization. The content and extent of research and integration activities are derived from the requirements of utilization, which is determined by the demands of political and practical allies.

The functions of research, integration and utilization constitute different subtasks of knowledge transfer, but they are not necessarily sequential. They alternate constantly, particularly research and integration. This repeated change is shown in the production line. The production line ends at the time of utilization, the active use of science-based products by different actors. The different subtasks of RIU are described in more detail below:

2.2.1 Research

Research is a process that produces a specific form of knowledge by using scientifically accepted principles, methods and standards. (Schnell et al. 2008, p. 7 ff; Kromrey 2000, p. 67 ff.; Atteslander 1995, p. 30 ff.)

The main function of research for scientific knowledge transfer is to guarantee state-of-the-art scientific theories, methods and data that have been previously selected by integration activities as being relevant to the solutions of practical problems. In science there can be a lot of research that is devoted to theory development, and that follows the current trends in various scientific disciplines. Such research is normally independent from practical needs and influences. The process of integration reviews scientific research, identifies the type of research that is relevant for the solution of practical problems and separates it from the research that is not currently needed in practice, since there may be a lot of science that is irrelevant to current practical problem solutions.

The research process consists of the following steps (Bortz and Döring 2009; Schnell et al. 2008; Kromrey 2000; Atteslander 1995):

1. Identification of research questions
2. Definition of research objectives
3. Development of research methods
4. Data collection and data analysis
5. Scientific interpretation

At each of these research steps the scientific information has to be up to date. The underlying theories and research methods have to represent the most current information from the respective scientific discipline. Also, the criteria of "good scientific practice" have to be fulfilled (DFG 1998).

2.2.2 Integration

Integration is the direction of research to a practical problem with the aim of describing the problem and successfully solving it. This orientation is bi-directional: On the one hand, the demand of practice for science-based solutions has to be investigated and used for the selection of research questions, on the other, scientific research results are selected according to their relevance to the practical solution.

Integration connects scientific knowledge to actors' expectations in practice. Such a connection requires constant exchange between research and integration, and it is only possible with extensive knowledge of both science and political practice. Integration searches for matches, contradictions and gaps in practical goals. Congruence with goals serves as a signal for the successful description and solution of problems.

Gaps lead to adaptation, either by formulating new research questions or by learning from practice. If there are contradictions there is either an alignment of practical expectations with the actual available scientific knowledge, or scientific knowledge without practical relevance is abandoned.

Integration connects research with practical demands by selecting individual "bricks of knowledge." These bricks are valid for the purpose of integration because they reflect state-of-the-art science and because they are relevant in practice (Böcher and Krott 2010, p. 38; Jasanoff 1990, p. 250). The resulting products are integrative in that they meet the standards of scientific criteria and also deliver practical solutions. Integration, then, reflects the standards of both scientific and practical problem solving. However, integration is forced to compromise on both counts: In order to find practical solutions it reduces the amount of scientific knowledge weighed, and it also corrects the political expectations regarding existing scientific knowledge. Therefore, integration inevitably leads to friction with those who pursue broader research objectives as well as with those who represent the wishes of political actors.

2.2.3 *Utilization*

Utilization is the use of scientific knowledge by actors in practice. Actors decide by themselves what kind of knowledge and how much of it they want to use, and they can also force other actors to use this scientific knowledge. Utilization also includes the use of the scientific knowledge within the scientific discourse.

Utilization starts with the delivery of a finalized knowledge transfer product [scientific policy advice] to political and practical actors. The product can be delivered in different forms, such as written reports, brochures, suggestions for new regulation schemes and standards, or implementation guides (Böcher and Krott 2010, p. 41). Actors involved in practice (citizens, consumers, producers, associations, media, administration, politicians) then use scientific advice actively to solve problems.

Other potential products are disseminated to the scientific community via scientific papers or contributions to scientific meetings. This dissemination guarantees that the scientific research that is used to advise policy makers reflects the most current scientific knowledge and that it ensures the research quality of the entire RIU process.

Utilization leads to changes in certain actors' behaviour. It needs its own procedures and resources to gain influence on the actions of other actors. In utilization, knowledge transfer is no longer in the hands of the scientists involved. Once a knowledge-transfer product is released, scientists can no longer influence its content and use. Making a clear distinction between integration and utilization is very important. Integration is often a form of internal (hypothetical) information processing and mutual adjustment, be it solely in the minds of researchers or by means of collaboration with practitioners. The aim of integration is the development of knowledge transfer products of high practical relevance that can be released for the use of political actors, provided that they are finalized after a continuous bi-directional process between integration and research activities.

2.3 The Production Lines

The production lines illustrate and quantify the entire knowledge transfer process. They help to visualize the alternation between activities of integration and research until a utilization product is released.

The production lines enable us to compare the amount of time invested in research, integration and utilization. Furthermore, the activities of research, integration and utilization can be analyzed and revealed using the concept of production lines (Böcher and Krott 2010: 43 ff). The lines begin within the sphere of integration (e.g., a research project that is commissioned by a ministry) and lead back into the sphere of research (e.g., the demand of the ministry for science-based advice is reflected in the creation of a research project). The production lines later return to the sphere of integration (e.g., preliminary research results and their political implications are discussed with the ministry). This process continues until the finished advisory product can be delivered to an external target group. Switching between the subtasks of research and integration is typical of such processes. These production lines between research and integration cover the usual internal procedures of the scientific knowledge transfer processes. The lines reveal that a single advisory product generally combines integration and research and that it takes time to finalize a scientific knowledge transfer product (Böcher and Krott 2010: 25) (Fig. 2.1). The lines continue within the utilization process. Different actors use the scientific advice as a basis for their activities. They can also force other actors to consider and use the scientific advice. The periods of utilization can differ strongly. Some scientific advice is used for only months; other scientific results can be used for hundreds of years.

The production line allows the separation of the time resources invested and the quality of the activities. Even if research, integration and utilization are short, they can meet high quality standards and, conversely, a long period of research does not guarantee a high standard.

2.4 Allies for a Successful Transfer of Knowledge

The RIU model emphasizes the relationship between the criteria for successful knowledge transfer and the specific actors serving as allies for knowledge transfer. We speak of four different roles an ally can play: internal, external, learning and wise ally. Internal and external allies differ in the degree of involvement they show the research process. Learning and wise allies depend on the decision process of the actor itself. In practice, an ally can fulfil more than one role simultaneously, e.g. an internal ally that is also a learning ally.

> Allies are all actors that support knowledge transfer from science into political practice by means of their power.

2.4.1 Transfer Through Internal Allies

A mechanism that helps to transfer scientific knowledge into practical action is the political production, by internal allies, of acceptance. Internal allies are actors who are actively involved in the research process and the utilization of its results. They adopt a scientific solution because it promises them advantages over other actors. Internal allies use their available resources to ensure that other actors also adopt the solution. It is important to know the reasons why such actors want to follow a science-based solution and whether they can promote its acceptance by other actors. Weak actors do not help much in knowledge transfer, even if they are true followers of scientific advice, because they lack the power to exert enough of an influence over other actors so as to have them accept the advice as well.

2.4.2 Transfer Through External Allies

Another mechanism is the enforcement of science-based solutions by external allies. External allies are actors who do not participate in the research project, but who have the power to exert external pressure on other actors so that the latter seek cooperation with science and adopt science-based solutions. A strong actor forces other actors to solve their problems by themselves, supported by using scientific advice. This could reflect the strong actor's interests: If the other actors can solve their problems better by making use of scientific expertise, the strong actor does not have to invest resources to search for solutions.

2.4.3 Transfer Through Learning Allies

From the point of view of the actor, there is the option of learning. Learning allies are those actors who learn something new from scientific research and its results, in terms of their own interests. In fact, political actors assess scientific expertise always in light of their own interests, and seek to benefit. Nevertheless, they can sometimes alter the interpretation of what their own interests are, based on scientific fact. Actors would then change their own positions and their actions.

An example is flooding. If an actor wants to invest money to build houses near a river but scientists deliver strong empirical data indicating that during the next couple of years flood problems will arise, then actors might change their behaviour based on science. Actors' self-interests lead to their learning and help them accept critical scientific results. Guided by their interests, actors deal with scientific findings and seek their advantages. Based on scientific evidence, actors can see their interests in a different light, and then they reconsider their positions.

Critical and empirical science, the kind that acts independently from the direct influence of practice, is an important precondition for learning allies, because only independent research can produce critical results that contradict and challenge the expectations of practice. The production of solutions to issues of sustainability and the environment cannot function without sufficient scientific freedom.

It is possible that potential threats to actors' interests could trigger rethinking. To achieve such an effect, hard scientific facts and hypotheses are necessary. Learning allies exist, but political actors learning from scientific insight is a rarely-seen effect. Environmental issues cannot rely exclusively on the existence and the actions of learning allies.

2.4.4 Transfer Through Wise Allies

From the perspective of practitioners, scientific results reflect just one type of information. Most practical decisions are also based on traditions, sector belief systems, everyday knowledge and practical experience. Practical actors who have to make decisions are faced with the difficult task of evaluating various types of information and forming an overall judgement on which to base their decisions. Science provides little assistance with this task, since it lacks the ability to judge forms of knowledge other than scientific reasoning. Science only has the potential to reflect scientific thinking critically, not the many other types of knowledge that may be of worth.

Wise allies are those actors who are able to assess the importance of scientific knowledge against all other information correctly and to place scientific knowledge correctly within the overall decision-making process. They follow rules that consider more than solely scientific criteria. Science normally plays only a limited role in the generation of all the information used to make decisions. The idea of wise allies is very important for environmental issues. Environmental research results rely often on implementation by wise allies. For these allies, science plays only a limited role in decision-making. They make decisions based on different knowledge sources, of which science is a part, and by no means the only guiding criterion.

2.5 Transfer Through Integrators

All allies use scientific knowledge in a way such that it serves their own interests, and not only because of the objectives and epistemic aspects of science itself. This tension might not satisfy all scientists who are interested in the practical implications of their work. Nevertheless, individual interests are the reason why scientific knowledge can be incorporated into practical actions and political decision-making. The functions of the different allies for scientific knowledge transfer show that science may have an epistemic influence on practical decisions even in conditions under which political and practical actors follow their self-interests. This is an important difference between the RIU model and certain narrow functional models of scientific knowledge transfer that are based on simple public choice considerations and in which science does not have any epistemic influence on decision-making processes (see Chap. 4).

However, one must recognize that there is a tension between political interests and the cognitive interests of science. Political issues threaten the integrity and cognitive aspects of science. We have seen above that integration requires both an understanding of scientific principles and of practical aspects and actors' interests. Therefore, for successful scientific knowledge transfer by integration scientists have to be open-minded to both areas, scientific knowledge (which should be easy for them) and the requirements of the actors' practice. Scientists themselves need not necessarily adopt this bi-directional role. Rather, professional integrators can facilitate the integration tasks. They must build a close and trustful relationship with scientists and they have to understand the principles of scientific research at least to a degree. These integrators also have to build trust with practical actors. They have to understand the specific demands of practice, the time constraints and the interests on which practical actors base their decisions. The role of the integrator is to mediate between science and practice and to complete integration tasks. These integration tasks can overburden scientists, and they can lose their scientific legitimacy if they come too close to practical interests. There is a danger that scientists may not be able to ensure the quality of their work if they concentrate too much on doing integration—it simply takes time and lies outside the core of every scientist's competence. Developing professional integrating services might be a challenge for consultants.

2.6 Bricks of Successful Scientific Knowledge Transfer in the RIU Model

Processes of scientific knowledge transfer that influence (political) practice successfully have to incorporate research, integration, and utilization as individual activities that have to be carried out at the highest possible standards. Activities of the RIU model have to be combined in specific production lines. Within research, integration, and utilization, in turn, different subtasks also have to be fulfilled. These subtasks prescribe criteria for the description and explanation of procedures of successful scientific knowledge transfer. The main criteria for the three different aspects of the RIU model are:

1. Is there high quality scientific research?
2. Is there pluralistic and foresighted integration?
3. Is there assurance of a democratic utilization?

To assess and further elaborate upon these critical questions, research, integration, and utilization have to be understood in more detail.

2.6.1 High Quality Scientific Research

High quality scientific research consists of the following subtasks:

- Assessing current scientific information
- Compliance with the procedures of good scientific practice
- Cooperation with other scientific institutions and projects
- Independent meaningfulness of scientific findings

The criteria for research are depicted from the science system. They reflect the quality and the latest development stages of underlying scientific information and its production and use in the research process, by researchers themselves or by scientific institutions that further elaborate and use scientific information. Generally speaking, the research has to be an important part of the scientific discourse in order to maintain high scientific quality.

2.6.1.1 Assessing Current Scientific Information

Research that informs scientific knowledge transfer processes in respect of current political and practical problems has to represent the latest stage of development in the respective field. For instance, research that informs international climate policy should represent the newest and best possible scientific climate scenarios and prognoses that are currently available and that are evaluated by the scientific community in peer review processes. Part of the research process is to assess the national and international state of the art of research, to identify possible knowledge gaps and scientific inconsistencies or controversies, and to identify unsolved research questions. Research process activities that support these tasks are, for example, literature or data research, or contacting other scientists in the same research area. These activities help identify knowledge gaps to ensure that the scientific information that is produced and used is state-of-the-art.

The important questions to assess the quality of incorporated scientific information are:

- *Has the national and international scientific literature been collected and reviewed?*
- *Have the national and international data sources been collected and reviewed?*

2.6.1.2 Compliance with the Procedures of Good Scientific Practice

In times of scientific scandals and public discussions about scientific misconduct, the compliance of scientific research with procedures of good scientific practice is crucial. First, only science and scientific results that reflect good scientific practice can legitimately claim to become the foundation of political and practical science-based decision-making. Second, science risks losing its legitimacy in society if non-compliance with good scientific practice becomes apparent. Practical actors might lose trust in scientific work and its research results if there is any suspicion of violation of procedures of good scientific practice. The IPCC lost some of its public trust when it became apparent that some of the data it used were not secured by procedures of good scientific practice, like professional peer review processes (Beck 2011, 2012). There are established criteria available for the assessment of good scientific practice, for example, the criteria of the German Research Foundation and the UK Director General of Research Councils (DGRC) and Chief Executives of the Research Councils (DGRC and Chief Executives of the Research Councils 1998; DFG 1998).

The following are important questions to assess scientific quality according to standards of good scientific practice (DFG 1998: 7):

- *Does the scientific work adhere to professional standards?*
- *Are the research process and its results clearly documented?*

- *Will new findings be scientifically published?*
- *Has the expertise of external researchers been called in when facing difficult questions?*

2.6.1.3 Cooperation with Other Scientific Institutions and Projects

Cooperation between scientific projects and institutions (universities, private or public independent research institutes) is important in order to make use of the synergy that may result, and to maximise the potential for innovative insight. In addition, early cooperation can help avoid overlapping research and can increase the efficiency and effectiveness of research funding or of financial resources in general. The exchange of scientific information and scientific cooperation with research institutions at local, regional, national, and international levels makes it easier to detect knowledge gaps and to establish strategic cooperation in research working.

Some relevant questions to help assess cooperation with other scientific projects and institutions are:

- *Are other relevant and current scientific projects known and has the potential for cooperation been evaluated?*
- *Is there scientific exchange with relevant research institutions at local, national, and international levels?*

2.6.1.4 Independent Meaningfulness of Scientific Findings

In many big research projects consisting of a lot of smaller, individual research projects, a synthesis of results is mandatory. But this synthesis does not help scientific knowledge transfer in political practice. Research syntheses risk weakening important results of individual research projects, losing these results, or generating a lack of empirical descriptions of the individual parts of the entire research project. It is therefore important for the quality of research that individual research findings should be independently informative and empirically substantiated.

If this is the case, the quality of research carried out by larger research groups increases with regard to overarching research questions. The question used to assess the meaningfulness of independent scientific findings is:

- *Are scientific findings independently informative and empirically substantiated?*

2.6.2 Pluralistic Foresighted Integration

Pluralistic foresighted integration consists of the following subtasks:

- Orientation toward public goals (like sustainability, …)
- Relevance in regard to political processes
- Relevance in regard to allies
- Target-group oriented intermediation

Within integration activities, research is directed towards practical and political questions.

First, this means that the normative goals of society or of the actors demanding science-based information must come into play in integration.

Second, integration is an activity that does not necessarily need to be carried out by scientists themselves. Rather, special institutions, e.g., departmental research institutions, can perform important integration tasks, as can single actors, like consultants, who are able to understand scientific research results and the demands of practitioners who are involved in integration activities. Scientists can serve as integrators as well, but in such cases they alternate between the roles of researcher and integrator.

Third, integration implies the existence of some kind of foresight, since the integrating actors often estimate current or future demands of practical actors for science-based knowledge. Therefore we can speak of "foresighted" integration.

Fourth, integration reflects the plurality of society's interests. In modern societies there are many actors demanding science-based information. As mentioned above, in such cases these actors are "allies" of scientific knowledge transfer because they are able to implement science-based knowledge within their realm, even against resistance, due to their power. At first glance, this would seem to be a problem for democratic processes, and a situation that could lead to a "biased" use of scientific information, in a way that powerful actors often prefer. These risks remain small as long as the criteria for scientific knowledge transfer are fulfilled, in spite of the presence of powerful actors using science-based information.

An important factor to consider is that, in a pluralistic society, there are many individual power relationships in which some actors are able to influence others by using their power resources to implement scientific research results. In addition, many such relationships can change over time, if formerly weak actors become powerful, creating new opportunities for knowledge transfer. Such an example is the use of nuclear power in Germany: In the 1970s, as a result of extensive discussion on the potential dangers of atomic energy, environmental groups and "alternative" scientists wanted the German government to stop using nuclear energy. But it was not until the 2000s that—due to the great nuclear catastrophe in Fukushima (Japan)—a policy window opened that enabled these formerly weak actors to become more powerful and to force other actors to accept alternative science-based information. Therefore, within classic pluralistic theories we argue that many power relationships exist at microcosmic levels within society and can change over time. This multiplicity of power relations offers a variety of options for scientific knowledge to enter society and politics. The widespread idea of a society controlled in great measure by a few powerful actors ignores the fragmented character of power relations. Scientific knowledge does not need the alliance of the most powerful actor, because whenever one actor is slightly more powerful than another the first can push scientific knowledge through so as for it to become relevant for the second, weaker, actor.

2.6.2.1 Orientation Toward Public Goals (Like Sustainability, …)

Generally, in integration, research is directed towards society's political and practical questions. Our criteria for integration make it possible to assess the degree to which research reflects current political problems and practical demands. They help determine whether science-based information is being produced that really can help to achieve normative public goals, like sustainability.

Integration activities lead to the direction of scientific research toward public goals like sustainability. This means that the research process and its results are oriented toward ecological, economic, and social sustainability goals. It is important that existing deficits of research and its results regarding public goals are clearly expressed. Furthermore, it is necessary to clearly define public goals like sustainability and to develop them further, normatively. Knowing what specific public goals a research project follows and how it interprets them is decisive in order to assess its quality: In the case of sustainability research, it makes a big difference whether research is oriented toward "strong sustainability" or toward concepts of "weak sustainability" (Böcher and Krott 2011: 36 ff. or Böcher and Krott 2012). Therefore, the normative public goals, and their conceptualization, that are the basis of the research process, have to be clearly defined before the research process can be assessed with regard to integration tasks. In many cases, a democratically legitimized institution, like a ministry, supports research and defines the public goals to be achieved with the support of science-based information. In such cases, like in national research programmes for sustainability, the funding institution defines overarching public goals that have to be pursued by the individual research projects.

It is also important for scientific findings to be interpreted in the context of potentially different and competing public goals from various institutions or from one, and at national and international political levels. In many cases competing, or even contradictory, public goals exist—like aspirations of economic prosperity that may contradict goals of strong sustainability. Within integration such contradictions should be revealed in order to establish realistic expectations for the success of scientific information with regard to specific competing public goals.

Another important aspect is that researchers have to be independent from public funding authorities in order to able to hold their ground against criticism of their research results. They have to be able to produce research results even if these are not politically opportune. They should be able to defend their positions against critics even if a ministry finances them. An example for this is the German Federal

Agency for the Environment. In many cases, the agency demands environmental policy measures that go against the political interests of the German Ministry for the Environment (BMUB). The agency supported a maximum speed limit in Germany, against the vote of the German government and, in the 1990s, the agency wanted to adopt an ecological tax reform that was in disagreement with the official political direction of the government (Böcher and Krott 2010).

Researchers' independence from private interests is also enormously important. In controversial issues, like sustainability or climate change, it is very important to have research institutions with public resources that allow independent, unbiased research. Among the important functions of independent research are to look out for the public interest and to expose interest-driven biased research. A lot of research having to do with environmental issues is privately financed (by big companies) and needs to be critically re-assessed by independent researchers that serve as watchdogs. The emergence of "critical" counter-expertise has been discussed in science for a long time (see for example Eden 1996; Bäckstrand 2004; Topçu 2008). The example of the use of nuclear power, described above, illustrates this well, as do others having to do with important current topics, like genetic engineering for food production. In this case, there are strong accusations of bias, in that much of the research in this field is financed by the food industry and needs to be critically re-assessed by independent experts.

Questions to assess orientation toward public goals like sustainability are:

- *Are the research process and its results orientated towards ecological, economic, and social aspects of sustainability?*
- *Are deficits regarding sustainability requirements clearly expressed?*
- *Are the values of sustainability clearly formulated, or have they been developed further according to norms, where needed?*
- *Is there a legitimate public contracting authority, like a ministry, that defines public interest goals?*
- *Are scientific findings coordinated in respect of the competing public interests of different public institutions at national and international levels?*
- *Are the researchers independent enough to hold their ground against criticism of their research results?*

2.6.2.2 Relevance in Regard to Political Processes

Research has to produce knowledge of relevance for political and practical actors. Research results should deliver analyses and explanations of current practical problems and should supply alternative solutions. Research should also deliver science-based prognoses about the future development of problems. In addition, political and practical actors need science-based information that is timely, so that they have enough time to react and make changes. Having a "stock of scientific knowledge" with regard to problems that might occur in the future is also important.

Scientific explanations should be directed to natural scientific and socio-economic causes of current problems (explanation) and to deliver solution strategies (configuration). Furthermore, an assessment of future problems should address relevant problems (prognoses). Having an orientation toward sustainability issues that have potential future relevance is an important strategy that buys time if there are problems demanding practical science-based information in the future.

There is a close connection between the relevance of science-based information and the reputation for seriousness and credibility of researchers or research institutions. If scientific institutions and the researchers working within are regarded as credible, their credibility and the relevance of their science-based advice remain even in times of public conflict about the choice of right solutions. In these cases, their credibility and the weight of their scientific arguments increase among their supporters and also among critics of their scientific positions.

If researchers have practical experience, for instance, with administration and politics, the relevance of their research results increases as well. This leads to a better understanding of the political process, of the type of information political actors need, and also of the types that are needed for intermediation.

Questions for the assessment of the relevance in regard to political processes are:

- *Do the research results deliver explanations for practical problems?*
- *Are there science-based alternative solutions for practical problems?*
- *Are there science-based prognoses about the future development of problems?*
- *Are scientific results being delivered to political actors in time for them to take action?*
- *Is there a stock of scientific knowledge about practical problems that could arise in the future?*
- *Do the researchers and institutions maintain their seriousness and credibility independently of current disputes?*
- *Do the researchers have practical experience with administration and politics?*

2.6.2.3 Relevance in Regard to Allies

Integration in support of science that leads to practical solutions has to incorporate the potential of concrete actors to become allies for scientific knowledge transfer. Science-based solutions have no practical impact as long as they lack a direct connection to concrete actors and their resources (power and finances). For each specific topic the relevant actors have to be identified and supplied by using individual intermediation strategies. The chance for the implementation of science-based solutions emerges only if the solutions are compatible with the interests, potentials for action and available resources of national and international relevant actors and institutions. This implies the identification of potential supporters and opponents of a certain science-based solution and the design of science-based solutions in a way such that it can attract enough allies. Detailed knowledge about the background of practical and political problems is important to develop solutions that are relevant for utilization. The development of win-win-strategies builds upon a realistic estimation of the interests of actors.

It is important to find different potential allies (internal, external, learning, and wise) and to incorporate these actors with their characteristics and potentials to the application of science-based information. These allies can be internal allies who are part of the research process and who can push the results through in pursuit of their own interests. An example of this would be a company that is part of a research process on climate adaptation strategies in economic enterprises and that is able to implement internally a business adaptation strategy that becomes mandatory for all workers and managers. Then there are external allies who are external to the research process but who are powerful enough to exert political pressure on others to use research results. An example of this type of ally would be a ministry that funds research on environmental issues and that has the power to adopt science-based solutions in a new environmental regulation scheme that every actor has then to follow. Learning allies are those who change their behaviour due to the emergence of new research results that affect their individual interests. Learning actors can be actors that reassess certain problems after new research results have been revealed that contradict their interests. For example, a house owner may learn through science that, due to climate change, floods are more likely to damage his building, which is near a river, and act accordingly. Wise allies are those who judge scientific findings encountered in their everyday reality and then make use of these in combination with other, alternative, forms of knowledge (like traditional or lay knowledge).

Questions for the assessment of relevance in regard to allies are:

- *Are there (internal) allies who participate in the research process and who push through the results in their own interest, even against political resistance?*
- *Are there (external) allies who can exert political pressure for the practitioners to cooperate with researchers constructively?*
- *Are there (learning) allies who are prepared to rethink old convictions in order to learn from scientific findings?*
- *Are there (wise) allies who judge scientific findings fairly from their everyday reality, and who make use of them?*

2.6.2.4 Target-group Oriented Intermediation

A group of quality criteria for integration addresses the intermediation of science-based information. Actors here are seen as target groups for science-based information and they need to be addressed by using individually specified products, media, and communication channels. Using a target group-orientated language and timely intermediation are further criteria for this group.

Questions for the assessment of target-group oriented intermediation are:

- *Are the research results presented in a language that is understandable for the target groups?*
- *Are those media used that reach the target group?*
- *Is the scientific information given at a time at which the practical discourse is open?*

2.6.3 Democratic Utilization

Subtasks of democratic utilization are:

- Contribution to democracy
- Contribution to rule of law (constitutional legality)
- Contribution to "good governance"
- Appropriate solutions to problems
- Participation in the scientific discourse

Utilization means the use of science-based information by practical and political actors, like politicians, stakeholders, citizens or the media. In addition, researchers use research results to contribute to the scientific discourse by producing publications or by giving academic speeches. Utilization, like the other activities of the RIU model, has to fulfil certain criteria in order to become part of science-based political or science-based practical action. The activities of utilization should meet high democratic standards, comply with the rule of law, and contribute to good governance, problem solutions and participation in the scientific discourse.

2.6.3.1 Contribution to Democracy

The enlightenment of society is essential for the functioning of democracy. This applies especially to sustainability and environmental issues, since such topics are sometimes new and usually cannot be directly perceived by citizens. The lobbyists of opposing sides often instrumentalise existing uncertainties in policy debates. Science-based advice on sustainability issues is therefore an important contribution to the assurance of democratic quality. It is important for the citizens and the media to have access to scientific results. The research process and the conditions of research funding have to be transparent and traceable. This means that there are instruments and procedures to inform the citizens and the media. Furthermore, scientific advisory processes have to be transparent and well-documented. Transparency enables democratic control and surveillance of science by governments and parliaments as well as by media and citizens.

Questions to assess the contribution to democracy are:

- *Are the research processes transparent?*
- *Do citizens have access to research findings?*
- *Do the media have access to research findings?*

2.6.3.2 Contribution to Rule of Law (Constitutional Legality)

Good quality of utilization is safeguarded by application and the observation of fundamental principles of the rule of law. Science-based solutions have to be applied in accordance with the law and through an impartial administrative procedure. The application has to be economically efficient. This economic principle means that science-based solutions are effective and that there is an adequate cost-benefit ratio of the solution in regard to the expected outcome.

Questions to assess the contribution to rule of law are:

- *Are the solutions applied conforming to the law and through an impartial administrative procedure?*
- *Are the solutions legally watertight?*
- *Is implementation economically efficient?*

2.6.3.3 Contribution to "Good Governance"

Governance means that decisions are taken by involving actors from civil society, like NGOs or citizens, and private actors, like economic enterprises. "Good" governance needs political decisions to be transparent, accountable, and traceable. This can be achieved by providing all actors in society with the same ability to express their concerns and to shape political decisions actively. Another important contribution to "good governance" is for the use of science-based problem solutions to be ensured in the long run by laws or regulations that integrate new science-based information, for example.

Utilization's contribution to good governance can be assessed by means of the following questions:

- *Do citizens participate in the solution?*
- *Do NGOs participate in the solution?*
- *Do commercial enterprises participate in the solution?*
- *Do consumers participate in the solution?*
- *Is application secured in the long term?*

2.6.3.4 Appropriate Solutions to Problems

The appropriateness of problem solutions means that utilization measures reflect existing problems accurately. Science-based solutions should be in line with the current state of technology in (political) practice. Risks arising from the use of any particular solution should be made known, and should be accepted or rejected.

Questions to assess the appropriateness of science-based solutions are:

- *Does the solution reflect the current state of technology in practice?*
- *Is the distribution of the solutions' risks visible and acknowledged?*

2.6.3.5 Participation in the Scientific Discourse

For quality control, research needs to be part of the scientific community. This also applies to research that is part of the RIU model and leads to science-based utilization products. Therefore, researchers have to produce a critical mass of products for the scientific community to safeguard their connectivity to the scientific community. Such products can be scientific publications, scientific contributions to conferences or participation in networks or topic-related science organizations. Quality does not result from the quantity of such scientific products. Rather, the production lines of the RIU model lead to a critical mass of scientific products, one that is needed to connect with the scientific community in a specific field or topic. This can only be evaluated qualitatively; there is no exact quantity that can be determined *ex ante*.

The question to assess the participation in the scientific discourse is:

- *Are the solutions incorporated into the current scientific discourse?*

2.7 Checklist for Successful Scientific Knowledge Transfer

As a result of our RIU model, and based on the results of our empirical research (see the following chapters), we can derive the following checklist (Fig. 2.2). The checklist presents individual quality criteria related to the subtasks of scientific knowledge transfer that support the transfer of scientific knowledge into political practice. By applying these quality criteria there is a greater probability for scientific projects to have an impact on practice. The checklist helps considering the specific factors that are relevant for research, integration, and utilization (RIU). Practical utilization is more likely to succeed if the number of questions on the checklist that can be answered in the positive is greater.

		Success Factors
1	R	Is there high quality scientific research?
2	I	Is there pluralistic foresighted integration?
3	U	Is there assurance of a democratic utilization?

1 High quality scientific research

Assessing current scientific information

- ☐ Has the national and international scientific literature been collected and reviewed?
- ☐ Have the national and international data sources been collected and reviewed?

Compliance with the procedures of good scientific practice

- ☐ Does the scientific work adhere to professional standards?
- ☐ Are the research process and its results clearly documented?
- ☐ Will new findings be scientifically published?
- ☐ Has the expertise of external researchers been called in when facing difficult questions?

Cooperation with other scientific institutions and projects

- ☐ Are other relevant and current scientific projects known and has the potential for cooperation been evaluated?
- ☐ Is there scientific exchange with relevant research institutions at local, national, and international levels?

Independent meaningfulness of scientific findings

- ☐ Are scientific findings independently informative and empirically substantiated?

2 Pluralistic foresighted integration

Orientation toward public goals (like sustainability, ...)

- ☐ Are the research process and its results orientated towards ecological, economic, and social aspects of sustainability?
- ☐ Are deficits regarding sustainability requirements clearly expressed?
- ☐ Are the values of sustainability clearly formulated, or have they been developed further according to norms, where needed?
- ☐ Is there a legitimate public contracting authority, like a Ministry, that defines public interest goals?

Fig. 2.2 Checklist for successful scientific knowledge transfer

- ☐ Are scientific findings coordinated in respect of the competing public interests of different public institutions at national and international levels?
- ☐ Are the researchers independent enough to hold their ground against criticism of their research results?

Relevance in regard to political processes

- ☐ Do the research results deliver explanations for practical problems?
- ☐ Are there science-based alternative solutions for practical problems?
- ☐ Are there science-based prognoses about the future development of problems?
- ☐ Are scientific results being delivered to political actors in time for them to take action?
- ☐ Is there a stock of scientific knowledge about practical problems that could arise in the future?
- ☐ Do the researchers and institutions maintain their seriousness and credibility independently of current disputes?
- ☐ Do the researchers have practical experience with administration and politics?

Relevance in regard to allies

- ☐ Are there (internal) allies who participate in the research process and who push through the results in their own interest, even against political resistance?
- ☐ Are there (external) allies who can exert political pressure for the practitioners to cooperate with researchers constructively?
- ☐ Are there (learning) allies who are prepared to rethink old convictions in order to learn from scientific findings?
- ☐ Are there (wise) allies who judge scientific findings fairly from their everyday reality, and who make use of them?

Target-group oriented intermediation

- ☐ Are the research results presented in a language that is understandable for the target groups?
- ☐ Are those media used that reach the target group?
- ☐ Is the scientific information given at a time at which the practical discourse is open?

3 Democratic Utilization

Contribution to democracy

- ☐ Are the research processes transparent?
- ☐ Do citizens have access to research findings?
- ☐ Do the media have access to research findings?

Fig. 2.2 (continued)

- **Contribution to rule of law (constitutional legality)**
  - ☐ Are the solutions applied conforming to the law and through an impartial administrative procedure?
  - ☐ Are the solutions legally watertight?
  - ☐ Is implementation economically efficient?
- **Contribution to "good governance"**
  - ☐ Do citizens participate in the solution?
  - ☐ Do NGOs participate in the solution?
  - ☐ Do commercial enterprises participate in the solution?
  - ☐ Do consumers participate in the solution?
  - ☐ Is application secured in the long term?
- **Appropriate solutions to problems**
  - ☐ Does the solution reflect the current state of technology in practice?
  - ☐ Is the distribution of the solutions' risks visible and acknowledged?
- **Participation in the scientific discourse**
  - ☐ Are the solutions incorporated into the current scientific discourse?

Fig. 2.2 (continued)

2.8 Conclusion: The RIU Model as an Analytical Framework for Successful Scientific Knowledge Transfer

By using the RIU model described above, ongoing or planned scientific knowledge transfer processes and their parts can be scientifically analyzed and evaluated. Additionally, individual factors, like subtasks, can be derived which help researchers, research funding institutions, and integrators to steer their research process with regard to practical implications. By using the RIU model and the checklist it is possible regularly to check the success of the entire scientific knowledge transfer process. The RIU model reflects the current state of the art of research in political science (see Chap. 4) and has been successfully used as an analytical framework in many empirical cases. As a conclusion to this chapter, three main questions lead to analysis and improvement of the process of scientific knowledge transfer:

1. Is there **high quality scientific research**?
2. Is there **pluralistic and foresighted integration**?
3. Is there an **assurance of a democratic utilization**?

The answers can be found by using the criteria for research, integration, and utilization, described above, to identify the potential for improvement of the process and its subtasks. However, not all criteria have to be fulfilled in every process at the same time. Rather, these criteria deliver a framework of possibly important aspects of scientific knowledge transfer. In a concrete research project aiming at practical utilization, its topic, its target groups, and its production lines determine the specific criteria that have to be used in that particular case.

Research projects that become strong in all three aspects, research, integration and utilization, have a high probability of producing science-based knowledge that is used in practice.

References

Atteslander, P. (1995). Methoden der empirischen Sozialforschung (8., bearb. Aufl.. ed., Sammlung Göschen; 2100). Berlin {[u.a.]: de Gruyter.

Bäckstrand, K. (2004). Scientisation vs. civic expertise in environmental governance: Eco-feminist, eco-modern and post-modern responses. *Environmental Politics, 13*(4), 695–714.

Beck, S. (2011). Moving beyond the linear model of expertise? IPCC and the test of adaptation. *Regional Environmental Change, 11*(2), 297–306.

Beck, S. (2012). Between tribalism and trust: The IPCC under the. *Nature and Culture, 7*(2), 151–173.

Böcher, M., & Krott, M. (2010). Umsetzung des Konzepts einer modernen Ressortforschung im Geschäftsbereich des BMU (Vol. 39/2010, Umweltbundesamt-Texte). Dessau-Roßlau: Umweltbundesamt.

Böcher, M., & Krott, M. (2011). *Forschungsfeld "Raumbezogene Nachhaltigkeitsforschung" - Stand und Optionen für die Zukunft in Österreich*. Göttingen: unpublished Research Report.

Böcher, M., & Krott, M. (2012). Nachhaltigkeitsforschung in Österreich. *GAIA-Ecological Perspectives for Science and Society, 21*(2), 155–157.

Böcher, M., & Krott, M. (2014a). *Mit Wissen bewegen! Erfolgsfaktoren für Wissenstransfer in den Umweltwissenschaften*. München: Oekom.

Böcher, M., & Krott, M. (2014b). The RIU model as an analytical framework for scientific knowledge transfer—the case of the "decision support system for sustainable forest management within climate change". *Biodiversity and Conservation, 23*(14), 3641–3656.

Bortz, J., & Döring, N. (2009). *Forschungsmethoden und Evaluation : für Human- und Sozialwissenschaftler; mit 87 Tabellen (4., überarb* (Aufl ed.). Heidelberg: Springer-Medizin-Verl.

Cash, D. W., Clark, W. C., Alcock, F., Dickson, N., Eckley, N., & Jäger, J. (2002). Salience, credibility, legitimacy and boundaries: Linking research, assessment and decision making. KSG Faculty Research Working Paper 02-046. Cambridge MA: Kennedy School of Government, Harvard University.

DFG, (German Research Foundation). (1998). Safeguarding good scientific practice. *Memorandum*. Weinheim: Wiley-VCH.

DGRC and Chief Executives of the Research Councils. (1998). Safeguarding good scientific practice.

Dharmawan, B., Böcher, M., & Krott, M. (2016). Failure of science-based win-win solution in fishery management: Learnings from Segara Anakan Waters. Central Java, Indonesia. under review.

Eden, S. (1996). Public participation in environmental policy: Considering scientific, counter-scientific and non-scientific contributions. *Public Understanding of Science, 5*(3), 183–204.

Guston, D. H. (2001). Toward a "best practice" of constructing "serviceable truths". In M. Hisschemöller, R. Hoppe, W. N. Dunn & J. R. Ravetz (Eds.), *Knowledge, Power, and Participation in Environmental Policy Analysis. Policy Studies Review Annual, 12*, (pp. 97–118). New Brunswick/London: Transaction Publishers.

Heim, J., & Böcher, M. (2016). CITES and science: Using the RIU-model to analyze institutionalized scientific policy advice in Germany for the case of ivory trade. *Journal of International Wildlife Law and Policy* (forthcoming).

Hoppe, R. (1999). Policy analysis, science and politics: From 'speaking truth to power' to 'making sense together'. *Science and Public Policy, 26*(3), 201–210.

Jasanoff, S. (1990). *The fifth branch : Science advisers as policymakers*. Cambridge, Mass: Harvard University Press.

Kromrey, H. (2000). Empirische Sozialforschung : Modelle und Methoden der standardisierten Datenerhebung und Datenauswertung (9., korrigierte Aufl.. ed., Utb, 1040). Opladen: Leske und Budrich.

Krott, M. (2012). Value and risks of the use of analytical theory in science for forest policy. *Forest Policy and Economics, 16*, 35–42.

McNie, E. C. (2007). Reconciling the supply of scientific information with user demands: An analysis of the problem and review of the literature. *Environmental Science and Policy, 10*(1), 17–38.

Mitchell, R. B., Clark, W. C., Cash, D. W., & Alcock, D. (2004). Science, scientists, and the policy process: Lessons from Global environmental assessments for the Northwest forest plan. In K. Arabas, & J Bowersox (Eds.), *Forest futures. Science, politics, and policy for the next century* (pp. 95–111). Lanham, Boulder, New York, Toronto, Oxford: Rowman & Littlefield.

Moll, P., & Zander, U. (2006). *Managing the interface. From knowledge to action in global change and sustainability science*. München: oekom-Verl.

Müller-Rommel, F. (1984). Sozialwissenschaftliche Politik-Beratung - Probleme und Perspektiven. *Aus Politik und Zeitgeschichte, 32*(B25–84), 26–39.

Nagasaka, K., Böcher, M., & Krott, M. (2016). Science-policy interaction: Who does what? *Case of the forest and forestry revitalisation plan in Japan*. under review.

Renn, O. (2003). Sozialwissenschaftliche Politikberatung: Gesellschaftliche Anforderungen und gelebte Praxis. TA-Informationen(1), 4–13.

Sarewitz, D., & Pielke, R. A, Jr. (2007). The neglected heart of science policy: Reconciling supply of and demand for science. *Environmental Science and Policy, 10*, 5–16.

Schnell, R., Hill, P. B., & Esser, E. (2008). Methoden der empirischen Sozialforschung (8., unveränd. Aufl.. ed.). München {[u.a.]: Oldenbourg.

Science, G. O. F. (2010). *The government chief scientific adviser's guidelines on the use of scientific and engineering advice in policy making*. London: Department for Business, Innovation and Skills.

Stevanov, M., Böcher, M., Krott, M., Krajter, S., Vuletic, D., & Orlovic, S. (2013). The research, integration and utilization (RIU) model as analytical framework for the professionalization of departmental research organizations: Case studies of publicly funded forest research institutes in Serbia and Croatia. *Forest Policy and Economics, 37*, 20–28. doi:10.1016/j.forpol.2013.03.006.

Topçu, S. (2008). Confronting nuclear risks: Counter-expertise as politics within the French nuclear energy debate. *Nature and Culture, 3*(2), 225–245.

Ugolini, F., Massetti, L., Sanesi, G., & Pearlmutter, D. (2015). Knowledge transfer between stakeholders in the field of urban forestry and green infrastructure: Results of a European survey. *Land Use Policy, 49*, 365–381.

Weingart, P. (1999). Scientific expertise and political accountability: Paradoxes of science in politics. *Science and Public Policy, 26*(3), 151–161.

Chapter 3
Case Studies

3.1 Background and Methodology of the Case Studies in Austria

3.1.1 The Background: Austrian Sustainability Research Programmes

The case studies we have conducted, described below, were part of the research project, "Optimisation of Scientific Knowledge Transfer from the Environmental Sciences" (*Optimierung des Praxistransfers aus den Umweltwissenschaften*). This project, which ran from 2009 to 2013 and which the Austrian Ministry for Science supported as part of the sustainability research programme, proVISION, describes and explains, through selected examples, the impact that different research programmes related to sustainability have had on practical applications in Austria. The result was the book, in German, *Mit Wissen bewegen!* ("To move through knowledge!") (Böcher and Krott 2014), which, among other things, describes the new RIU model of scientific knowledge transfer theoretically, with the help of which the previously selected empirical cases were analysed. The RIU model presents factors related to the three important areas of knowledge transfer: research, integration and utilization, as described already in Chaps. 1 and 2. The investigation of knowledge transfer processes in Austrian sustainability research projects is methodologically justified in that Austria can be seen both within Europe and worldwide as a pioneering land for such projects, given that the country had already begun to establish special state-sponsored research projects for sustainability in the early 1990s (Böcher and Krott 2012).

This chapter has been translated from *Mit Wissen bewegen! Erfolgsfaktoren für Wissenstransfer in den Umweltwissenschaften* written by Michael Böcher and Max Krott.

M. Böcher and M. Krott, *Science Makes the World Go Round*,
DOI 10.1007/978-3-319-34079-1_3

Here, and among others, actor networks were to be supported that again exerted an influence on the determination of funding policy (Pesendorfer 2007, p. 171). As relates to the facet of transdisciplinary sustainability research, within these early research activities in the 1990s and the beginning of the 2000s the most outstanding was the Cultural Landscape Research (*Kulturlandschaftsforschung,* KLF) programme, which emerged after public consultation regarding a strategy paper emitted in 1992 by the Austrian Ministry for Science. This was sent to over 500 actors in the Austrian research community (Pesendorfer 2007, p. 171). The ways in which the concept of sustainability could be implemented regionally (spatially) were meant to be shown under the umbrella of the KLF (Pesendorfer 2007, p. 171). Through the programme, research results were meant to reach, in particular, a high degree of feasibility for research, policy, planning and education (knowledge transfer), and the applicability for legislation and administration had to be secured (Bundesministerium für Wissenschaft 1995, p. 39). At that time the KLF set a considerable emphasis of the research funding on the areas of the environment and sustainable development for Austria (Pesendorfer 2007, p. 172); it ran from 1995 to 2004 and had an allocation of approximately 17 million euros. This funding was provided jointly by the Federal Chancellery, the Ministries of the Environment, Agriculture and the Economy, and the federal states. In addition, international research funds, EU regional grant funds and private monies were also used (Begusch-Pfefferkorn 2006, p. 125).

The success of the KLF is estimated to be very high: Its "knowledge balance sheet" includes successes in the area of "scholarly gains" (numerous cited publications, even internationally, qualification of young scholars, interdisciplinary networking of participating scholars). In addition, it also achieved successes in internationalisation (networking of KLF results and international research). Furthermore, many practice-oriented effects could be observed (activities like artistic events, school projects, television, radio or similar programmes, and news articles covering KLF research). Along with these notable effects, the KLF knowledge balance sheet mentioned above also includes leverage effects. For example, additional resources were raised for projects; there were follow-up projects or successful knowledge transfer in specific non-partisan policies, successful promotion of women, innovative language work through an improvement in the language abilities of participants in terms of conciseness, clarity or tangibility. Numerous projects and project results were honoured with prizes or a particularly high demand for publications (Begusch-Pfefferkorn 2006, pp. 125–139).

Not only did the internal balance sheet show very positive results: so did the external evaluation. A summary said: "In an international comparison, the team of experts judges the KLF programme to be an unusually brave, creative, innovative and altogether successful undertaking" (Hüttl 2003, p. 8; own translation). It is not surprising that the experiences of the Austrian KLF were included in several scientific publications, including those of researchers outside Austria, for example, in a publication by Brand (2000). With the KLF, Austria was able to establish a "role model" for sustainability research that would broadcast internationally and from which other states, like Germany, could learn. In 2004 Austria began proVISION, the transdisciplinary sustainability research programme, which appeared financially slim when compared to the KLF. In contrast to the latter, it had a significantly smaller allocation of approximately 7.6 million euros (12.6 million euros were planned by 2006; for the KLF there were 17 million euros available) (Böcher and Krott 2012, p. 156), but it conformed to the path taken with the KLF, of critical, participatory and inter- and transdisciplinary sustainability research.

ProVISION—provision for nature and society, "concentrates on the subjects of climate change, spatial development, and sustainable ways of life, is dedicated to the ways in which society deals with ecosystems and their benefits and is in dialogue with society from the beginning" (Paula et al. 2004, p. 29; own translation). In a similar way as with the KLF, cooperation between science and practice and equal opportunity between men and women, among other things, are the focus. ProVISION was to form the scientific fundament of Austrian sustainability strategy. In doing this, proVISION promoted "scientific innovative projects that make significant contributions to methods, theories, quality criteria for inter- and transdisciplinary science for sustainable development and that support innovative action in practice" (BMWF 2007, p. 1; own translation). In this, proVISION was the programme closest to the ideal of transdisciplinary sustainability research.

On the basis of both of these innovative Austrian research programmes for sustainability, it becomes clear that Austria was at the vanguard in this very early on, in what regards sustainability research oriented towards knowledge transfer from the environmental sciences (Böcher and Krott 2012). An investigation that was part of a research project funded by the EU confirmed that, within the sustainability research carried out in Europe there were forerunner states that had already begun to implement the second generation of sustainability research programmes at the time of the investigation, whereas other countries were only beginning to develop a first generation of such programmes (Whitelegg 2002, 2006; Whitelegg et al. 2002). Beyond this and into the programme content and the actual beginning of the individual research programmes, Austria, with its KLF programme, which began as far back as 1995 and which followed inter- and transdisciplinary methods in terms of critical sustainability research, represents an innovative and pioneering country in Europe and a forerunner of the first generation of sustainability research. Its programmes radiated out to other countries, like Germany and Switzerland, which, in the development of their own programmes for sustainability research, learned from the experience of Austria (Hüttl 2003, p. 3; Böcher and Krott 2012, p. 156; Schneidewind and Singer-Brodowski 2013, pp. 202–205). Many of the projects promoted in the KLF and proVISION had the aspiration of leading to practical changes, be they changed laws and directives or behaviour modifications in the actors involved in practice.

Because of the significance of Austria for this specific type of research programme for sustainability, we chose this country for our case studies, because in it we expected to be able to examine particularly well the mechanisms involved in the successful transfer of knowledge from the environmental sciences. The individual case studies demonstrate the capability of the RIU model in terms of the description of detail in cases of successful knowledge transfer:

1. Hand in hand—Life 2014 in the Pinzgau region—Common regional development cooperation instead of parish-pump politics
2. Ski Tourism within climate change—the STRATEGE project
3. Indicators for human impact on the biosphere and sustainable development
4. Options for Viennese agriculture and horticulture
5. Establishing a legal basis: Implementing the EU Water Framework Directive in Austria
6. Research in the Neusiedler See—Seewinkel National Park region

These six cases cover a broad spectrum: There is an example in which there is an approximate manifestation of the rare linear model of knowledge transfer ("Implementing the EU Water Framework Directive in Austria"), and examples of close cooperation between science and local practice in transdisciplinary research ("STRATEGE" and "Life 2014 in the Pinzgau region"). In addition, there are cases that make clear that the desired success rates in terms of knowledge transfer are not reached in the short term. Rather, it takes time to be able to assess all facets of the possible effects of said transfer (as in the example of the study on "Indicators for human impact on the biosphere and sustainable development"). This can be used to show that successful knowledge transfer cannot be judged in the short run, but that instead it must also involve long-term ex post observation. Another variant provides evidence for the notion that, especially in sustainability research, it is important to couple research oriented toward the long term to existing contexts, in order to be able to investigate, for example, the relationship between humans and the environment and effects on biodiversity, with a view toward a local area (as in the example of the "Neusiedler See—Seewinkel National Park" study). Furthermore, the case studies make clear that knowledge transfer depends on those allies being determinant who have the power to push through solutions within their realm of influence. The detailed description of the role of different allies for knowledge transfer within the RIU model was possible empirically for the first time with this project by means of the selected cases (Table 3.1).

Table 3.1 Overview over the case studies' findings on allies

Case (funding programme)	Allies	Who? (main actors)
Hand in hand—Life 2014 in the Pinzgau region (KLF)	Internal	Local mayors and interns
Ski Tourism within climate change—the STRATEGE project (proVISION)	External	Professional Association of Austrian Cable Cars
Indicators for human impact on the biosphere and sustainable development (KLF, among others)	External	Austrian Parliament ("*Nationalrat*")
Options for Viennese agriculture and horticulture (KLF)	Internal/external	Various Magistrate Departments in Vienna City Councilwoman for Environmental affairs
Establishing a legal basis: Implementing the EU Water Framework Directive in Austria (KLF, among others)	External	EU, Austrian Ministry of Agriculture, Forestry, Environment and Water Management
Research in the Neusiedler See—Seewinkel National Park region (Basic financing from the state and financing of individual research projects from different programmes.)	Internal/external	National Park Administration Austrian Ministry of Agriculture, Forestry, Environment and Water Management State of Burgenland

3.1.2 Methodology

3.1.2.1 Development of a Theoretical Framework

First, the project developed the RIU model of scientific knowledge transfer on the basis of preliminary work and of an evaluation of existing literature on the requirements for successful knowledge transfer processes. This model served as a foundation for the rest of the surveys in Austria. This theoretical framework is presented in detail in Chaps. 1, 2 and 4 of this book.

3.1.2.2 Empirical Investigation

In the course of a meeting in Vienna with Austrian experts on sustainability research, and by way of telephone interviews, information was gathered in a broad survey about projects that could be the object of empirical case studies. In addition, the interviews were to deliver information about aspects of knowledge transfer that had not been yet considered in the development of the model and that would help to develop content questions for further surveys. With these surveys we could gather information on seven very promising projects, of which three ("Options for Viennese agriculture and horticulture", "Indicators for human impact on the biosphere and sustainable development", and "Research in the Neusiedler See—Seewinkel National Park region") found their way into the book in the end. The other projects were very similar in terms of the underlying mechanisms of knowledge transfer, or we were unable to gather enough information about them. In a first round of detailed empirical investigation in Austria (interviews with participating scientists and practitioners, as well as document analyses), data were collected on the course of the projects and on the practical effects.

Since the aim of the project was to show as good and wide a sample of various forms of knowledge transfer as possible, as well as thematically different projects, in a second round information was requested from all scientific project leaders participating in the KLF and proVISION sustainability research programmes about other positive examples of knowledge transfer. This was done by means of a written questionnaire via email. This questionnaire yielded many other references to yet unseen projects, and helped to estimate the quality of those seen until then: For example, the "Indicators…" project was mentioned repeatedly, and in the same way, several actors mentioned the "STRATEGE" project independently. Because of this information, some of the projects that had been shortlisted already were discarded again, and some that had not been considered were examined, e.g., the project on the EU Water Framework Directive. On the basis of these additional consultations with experts, four other projects were examined empirically in an intensive way: This included expert interviews with central key persons and an extensive evaluation of project documents (reports, publications, internet presentations, etc.).

The six case studies collected in this book are the result of these document analyses, written and verbal consultations with experts as well as interviews with persons directly involved in the projects.

3.2 Hand in Hand—Life 2014 in the Pinzgau Region—Common Regional Development Cooperation Instead of Parish-Pump Politics

3.2.1 Project Fact Sheet

Project

Transdisciplinary cooperation in university education. The "Life 2014" (*Leben 2014*) case study in the Hohe Tauern/Oberpinzgau (Austria) National Park region

Type of project

Teaching/research project: Students research a concrete practical case together with scientists who teach at various universities

Participating research institutions

- Institute of Landscape Development, Recreation and Conservation Planning (ILEN) of the University of Natural Resources and Life Sciences, Vienna (*Institut für Landschaftsentwicklung, Erholungs- und Naturschutzplanung* (*ILEN*) *der Universität für Bodenkultur Wien*)
- Division of Organic Farming (IFÖL) of the University of Natural Resources and Life Sciences, Vienna (*Institut für Ökologischen Landbau der Universität für Bodenkultur Wien*)
- Institute of Geography and Applied Geoinformatics of the University of Salzburg (*Institut für Geographie und Angewandte Geoinformatik der Universität Salzburg*)
- In cooperation with the Swiss Federal Institute of Technology in Zurich (ETHZ) and the University of Graz

Location

Oberpinzgau, State of Salzburg

Central question

What should landscape, land use and society in the Hohe Tauern National Park region look like in ten years?

Goals

- Designing future scenarios for succesful sustainable regional development in the Oberpinzgau region
- Drafting concrete implementation projects to achieve lively regional development

Timeline

2002-2005

Financing

- Cultivated Landscape Research (*Kulturlandschaftsforschung*, KLF), sponsored by the then Austrian Federal Ministry of Education, Science and Culture (Österreichisches Bundesministerium für Bildung, Wissenschaft und Kultur)
- Co-financing: Oberpinzgau Regional Association (*Regionalverband* RV 14)

3.2.2 From Parish-Pump Politics to Regional Cooperation

The idea of "parish-pump politics" is well known. Instead of engaging in problem solving at a community level, local politicians often think only of their own, local community, even if it is much more reasonable to cooperate with neighbouring communities and to pursue common regional interests. Regional cooperation is even advantageous, both for the individual community, and also for its partners in cooperation. Sustainable regional development needs cooperation among administrational borders.

The existence of "parish-pump politics" was the reason for the establishment of the project "Hand in hand—Life 2014 in the Pinzgau region" (Life 2014). "Life 2014" was to help develop concrete projects in regional sustainable development in nine local communities of the Austrian Oberpinzgau region in the federal state of Salzburg (northwestern Austria) (www.partizipation.at).

The project's most important steps were carried out in 2004, and the participating actors had to develop future scenarios with a 10 year perspective, therefore the year 2014 served as future reference point. Among the questions raised were those pertaining to the general possibilities for regional development during the ten years that would follow, the achievements that the region would see by 2014 and the ways in which the population would want to live and cooperate in its region (Glanzer et al. 2005, p. 1).

A total population of 22,000 lives in the nine communities of the Oberpinzgau region (Hohe Tauern National Park). All communities face similar problems that are typical for peripheral alpine regions (Glanzer and Muhar 2006, p. 6):

- Lack of traffic infrastructure: Access to the region by road is inadequate; the closest highway access is 1 h away.
- The role of agriculture and forestry as a main source of income for the local population is diminished
- Decreasing number of jobs for qualified people: There are no big industries in the region due to its peripheral location.

A very important economic factor for the region is tourism. In the winter season ski tourists come into the region, and during the summer holidays there are tourists who like to hike, climb, cycle or who have a great interest in the National Park.

However, the Oberpinzgau region is in no way a region for mass tourism, like other Austrian skiing regions or the neighbouring region of "Zell am See" (Glanzer et al. 2005, p. 1).

3.2.3 The Project: Life 2014

Life 2014 was a teaching/research project, which means that not only trained scientists were part of it. Rather, students from University of Natural Resources and

Life Sciences, Vienna (BOKU) and the University of Salzburg also participated in the research process. Here, they were meant to learn how to research one concrete topic during one semester. In cooperation with local citizens, the students would work on questions on the development of the region and would find solutions to particular problems (Glanzer et al. 2005, p. 1). This cooperation was meant to ensure that the wishes and problems of the local population could be integrated into the research process. The idea was to introduce a young and fresh perspective (Glanzer et al. 2005, p. 1). Many partners were involved in the "Life 2014" project: more than 40 students, more than 20 teaching staff, and more than 70 regional actors, among others, the mayors of the nine communities, the regional management of Pinzgau, and representatives of the Hohe Tauern National Park and the federal state of Salzburg.

3.2.4 Finding Topics for Regional Development

To begin with, central topics were identified that would centre around the main question of the ways in which landscapes, land use and the region's society would develop by the year 2014 (Glanzer et al. 2005, p. 3). Here, local citizens and external experts worked together (Glanzer et al. 2005, p. 3). By using different methods (interviews with mayors and the project partners, thematic workshops) the most important topics for regional development were identified (Glanzer et al. 2005, p. 3). A total 86 people worked together in the thematic workshop, where they formulated concrete questions for external experts and students to answer for local actors (Glanzer et al. 2005, p. 4). The most central problems were found within the topics of tourism, the future of the younger generation, health and social issues, traffic, regional cooperation, the identity of the Oberpinzgau region, the National Park and agriculture (Glanzer et al. 2005, p. 4).

Because of the Oberpinzgau region's long tradition of developing only sectoral problem solutions (e.g., separate solutions for agriculture and tourism that disregard potential synergies), the topics were integrated into so-called "polarity fields" (in a sense of main dichotomies that are central to the regional problems) (Muhar et al. 2006). The idea behind these was to better reflect the complexity of regional problems, i.e., far beyond unidirectional, sectoral aspects. The polarity fields would enable researchers to detect and counteract any contradictions in the formulation of a particular goal (Glanzer et al. 2005, p. 4). Subsequently, single working groups began to work on topics and questions comprising the individual polarity fields, meaning they had to deal with the inter-sectoral aspects of this material. Table 3.2 shows the polarity fields' topics and their main guiding questions.

Between 2003 and 2004 the students worked together with the working groups of the different polarity fields. They reviewed the current situation and developed scenarios for life in 2014. Local actors discussed and evaluated these in a public meeting. The next step was to develop concrete regional development projects. The results of the teaching/research project were then presented to all interested actors

Table 3.2 The polarity fields and its main guiding questions

Polarity field	Main guiding question
Single and together	"How can relations between communities be shaped in a way that benefits all?"
Fast and slow	"What will sustainable mobility concepts in the Oberpinzgau region look like in the year 2014"
Inside and outside	"How can external influences be combined with existing regional potential so as to achieve successful future regional development?"
Tradition and innovation	"How can new economic impulses be introduced between tradition and innovation?"
Young and old	"What social and leisure services does the Oberpinzgau region need to become optimally habitable?"
Wilderness and culture	"How can the National Park, agriculture, and tourism work together in the region?"

Source Glanzer et al. (2005, p. 5), Muhar et al. (2006, p. 203)

during a big public event in which more than 200 guests participated. In this project, the students worked in the Oberpinzgau region for 4 weeks, during which they were the guests of local citizens. The quality of the project and its recognition, even beyond regional borders, were confirmed with the "ÖGUT Umweltpreis" (an environmental award) in the category "participation for sustainability—innovation and local projects" (Böcher and Krott 2014, p. 27f).

3.2.5 *Main Effect on the Regional Association: From "Paper Tiger" to an Active Regional Actor*

What possibilities and what effects brought about by the project can be observed in reality today, in the year 2015? What important changes have taken place that are related to the project results?

A good example for a real impact resulting from the project can be found in the "single and together" polarity field. The situation in 2004 was one that many actors described as "parish-pump politics", which was typical for the region and which was an important factor hindering regional development (Glanzer et al. 2005, p. 11). Stronger cooperation between the nine communities was seen as the most important goal (Glanzer et al. 2005, p. 11). There had been a few examples of cooperation before, but none that comprised the whole Oberpinzgau region (Glanzer et al. 2005, p. 11). Among the reasons for this were rivalries between the nine communities, as well as differences in their budgetary margins (Glanzer et al. 2005, p. 11). In particular, budgetary problems led to the need for better cooperation between the communities (Glanzer et al. 2005, p. 11). From then on, cooperation was seen as an important way to reduce future costs and to deal with changing

Fig. 3.1 Explanation of the scenarios by the students during a workshop. *Source* Leben 2014 (*Photo*)

problems and duties, e.g., in society and in the school system (Glanzer et al. 2005, p. 11).

The students developed four scenarios that represented alternative developments by 2014 and presented them to the local citizens in the form of "storylines" (Glanzer et al. 2005, p. 16). Then citizens, students and teaching staff evaluated the scenarios (Glanzer et al. 2005, p. 16) (Fig. 3.1).

Table 3.3 shows the different core aspects of different scenarios in the polarity field "single and together".

These scenarios show a number of options that range from unification into one large community ("Griaß di!"), as the most extensive form of cooperation, to an unchanged situation in which every single community has to deal with its problems like before, and no cooperation exists ("nine against zero in Oberpinzgau"). However, in the evaluation process the actors involved argued for the scenario called "Hand in hand in one regional association". This scenario was considered to be innovative and realistic. It was also seen as an important step in the improvement of quality of life, and it was thought that it would lead to a regional development boom in the Oberpinzgau region (Glanzer et al. 2005, p. 17). "Hand in hand in one regional association" meant strengthening the regional association that has existed since 1996 (Glanzer et al. 2005, p. 17). The regional association has existed since

Table 3.3 Scenarios in the polarity field "single and together", own translation

Scenarios in the polarity field "single and together"	
Nine against zero in the Oberpinzgau	"In the year 2014 all communities are still autonomous and cooperate only sporadically"
"Griaß'di!"* in the Oberpinzgau community	"By the year 2014 the borders between the nine communities are dissolved"
Hand in hand in one regional association	"In the year 2014 the nine autonomous communities in the Oberpinzgau still exist. The 'Oberpinzgau regional association' is responsible for cooperation on topics concerning multiple communities"
Three administrative focal points	"By the year 2014 the nine communities of the Oberpinzgau are part of three administrative centres"

Source Glanzer et al. (2005, p. 16 f)
*"Griaß'di!" means "Hi!", it is a typical Austrian greeting

Salzburg laws on spatial planning made it mandatory, but it has only been a "paper tiger" and not very active.

Based on the "Hand in hand" scenario, the students developed concrete projects for the realisation of stronger cooperation between the nine communities. According to these, the existing "Regionalverband" Oberpinzgau (regional association Oberpinzgau) was to become more active and would develop a service centre with its own management (Glanzer et al. 2005, p. 32). The "Regionalverband" would organise central tasks that affect all nine communities, in order to relieve the communities financially. They could then use the financial resources saved to fund other tasks (Glanzer et al. 2005, p. 32). The newly established regional management of the association would represent the region, develop regional development projects that could be supported by European funding measures, and it would begin with the service centre planning (Glanzer et al. 2005, p. 32).

3.2.6 Active "Regionalverband" for Real Cooperation

Of the effects that allow us to determine that knowledge transfer was successful here, the most impressive is that, today, there exists an active "Regionalverband" (regional association) representing the nine communities of the "Oberpinzgau" region. This association has evolved step by step during the last few years in a way that had been suggested by the research project.

Another important detail is the employment of one manager for the Regionalverband. He is also the mayor of the "Krimml" community, and is responsible for the realisation of central projects. The regional association has been incorporated into the regional management for the whole Pinzgau region. Currently, there is institutionalised cooperation between the nine communities. The conclusion today is that "the region has left parish-pump politics and established a culture of cooperation that went beyond boundaries that seemed to be insurmountable."[1]

The Oberpinzgau region developed very well—not just because of the project "Life 2014" directly, but also in part due to its effects. Pinzgau regional manager Michael Payer points out that in the early 2000s, the Oberpinzgau region had the worst economic performance in the whole Pinzgau region, and this despite the fact that the "Hohe Tauern" National Park was the recipient of vast financial resources (personal communication, May 15, 2012). Today, according to regional manager Michael Payer the Oberpinzgau has developed in a way "nobody could ever dream of before" (personal communication, May 15, 2012). For instance, the population, the number of overnight stays and the number of jobs have grown during the last couple of years (Fig. 3.2). Today these numbers are well above the average for the whole Pinzgau region (Michael Payer, personal communication, May 15, 2012).

A very impressive example for an increased level of cooperation is the Oberpinzgau intermunicipal tax equalisation, which was part of the project's suggestions. Here, starting in late 2008, three communities agreed to pool all their tax revenues, which then have to be spent according to a common allocation formula. By now, all nine communities are part of this intermunicipal tax equalisation system. This helps to avoid parish-pump politics actively. If, for example, one company wants to relocate to the region, the communities do not have to compete with each other anymore, since every individual community receives tax revenues, regardless of the community the company chooses. In addition, all can participate in the selection of the ideal place. Intermunicipal tax equalisation serves as a compensation for the fact that not all communities are economically equal.

Now this concept serves as role model for all of Austria, and many visitors come to the region for information. The "Intermunicipal tax equalisation" project has won various prizes.

However, not only has strengthened cooperation between the communities relied on the influence of the "Life 2014" project. Concrete regional development projects were also among the students' results, and some of these have been implemented. At the end of the project, the regional association chose seven concrete project ideas for realisation (Table 3.4).

For example, the aim of the project "KunstVerjüngung" project was to create a regional monument by planting different tree species, to highlight the historical meaning of forests and to have an artistic silvicultural area to serve as a monument for visitors from the local population (Glanzer and Schauppenlehner 2006, p. 226). The partners involved in this project would be artists, public schools and the Austrian Federal Forests (*Österreichische Bundesforste*, ÖBF).

[1]Mayor Franz Nill (Uttendorf), cited from "Oberpinzgau revisited—Austausch zur ZwischenZeit 2009" http://dyn.boku.ac.at/pinzgau/index.php?nav_id=revisited, acccessed on: 27.10.2013, own translation.

Table 3.4 Project ideas that were recommended for realisation

Name of the project	Contents
KunstVerjüngung	Creating a scenic site by replanting a windthrown area with diverse tree species, according to an artist's guidelines
PinzGaul	Establishing a regional system for bicycle rental
VitalT	Establishing a system for providing health-related information to people over 60
akv-Die junge Werkstatt	Supporting the creativity of young people of ages between 15 and 25
Pferdewanderungen in der Nationalparkregion	Horseback riding in the National Park region to attract new kinds of tourists
Auf dem Weg zur Bioregion	Turning Pinzgau into a pilot region for sustainable development ("Bio-region")

Source Glanzer and Schauppenlehner (2006, p. 208 ff), own translation

Fig. 3.2 Touristic highlight in the Oberpinzgau region: the Krimml Waterfalls. *Source* Michael Böcher (*Photo*)

3.2.7 *Factors for the Success of Knowledge Transfer*

In the case of the "Life 2014" project, the following factors contributed to the success of scientific knowledge transfer.

3.2.7.1 Transfer Through Internal Allies

An important factor for the successful scientific knowledge transfer of "Life 2014" was cooperation with regional actors who have the power to actually implement the project's results in practice. Therefore, regions were sought beforehand in which there was a willingness to cooperate actively with the research project and where there was a greater likelihood for its results to be implemented. An important indicator for cooperation was the will of the regions to contribute with their own financial resources. The nine communities participated in the funding of the project by bringing in €47,000. This co-financing principle is also typical for European structural funds policy and prevents regions from applying for funding without any active financial role. The willingness to use its own funds is an indicator that a region is really interested in getting funded and wants to cooperate to achieve certain projects' goals (Böcher 2008). Integrating the financial means of the nine communities reflects the current practice in similar political programmes for regional development at the EU level.

The mayors were important internal allies for the project, as important representatives of the nine communities. They were involved in the research process and were very active in the utilization of its results. These key actors' involvement also helped convince other actors of the benefits of the research project. At the beginning of "Life 2014" the mayors were very sceptical. However, one step at a time they became important allies for the project, persuaded by the intermediary project results.

The importance of the function of these internal allies for scientific knowledge transfer can be illustrated by different projects that became reality: a stronger cooperation in the Oberpinzgau region, the employment of one manager responsible for networking, fund raising, and project development. In addition, the intermunicipal tax equalisation became reality, one that could have never been realised without strong allies. Marianne Auer-Enzinger, vice mayor of Niedernsill, states: "With the intermunicipal tax equality system for nine communities in Oberpinzgau (…) we created an absolute role model that is of high interest. The climate and cooperation between the communities are extraordinary (…). The aim of the tax equalisation system is to establish compensation between economically backward and strong communities."[2]

3.2.7.2 Research

In our RIU model of scientific knowledge, the knowledge transfer underlying research must be of high quality, and must consider the current state of scientific knowledge (reflecting the scientific state of the art).

---

[2]Source: Press release from 4 November 2011 www.niedernsill.spoe.at/presse.htm, accessed on 27 October 2013, own translation.

In the “Life 2014” project, appropriate social science research methods were applied (interviews with regional stakeholders, group discussions, surveys etc.). All methods used are documented in the final project report in the spirit of good scientific practice (Freyer and Muhar 2006, pp. 33–54). The teaching scientists’ experience and their supervision guaranteed that the scientific methods applied by the students were state-of-the-art.

The various theories (systems theory) and techniques applied (scenario technique, strengths-weaknesses analysis, project development) correspond to the most current methods in present research on the regional development of rural areas.

An indicator for the high quality of the scientific methods applied in the project is the discussion of the (intermediary) project results in the scientific community, and the existence of scientific publications. “Life 2014” led to numerous contributions for national and international symposia and for many scientific publications. As an example, one publication regarding the scenario technique used and the concept of polarity fields was published in GAIA (Muhar et al. 2006).

Due to its content and its main goal of sustainable regional development, the project also contributed to the programme objectives of the sustainability research programme “Cultural Landscape Research (KLF)”, of the Austrian Ministry of Science.

The project scientists cooperated with different partners and suppliers. According to Prof. Andreas Muhar of the University of Natural Resources and Life Sciences (BOKU) Vienna, this cooperation was initially a matter of applying transdisciplinary regional case studies in Austria that have been successfully tested before outside of Austria (personal communication, October 19, 2011). Therefore, there was a collaboration with ETH Zürich, which had already conducted similar training projects involving inter- and transdisciplinary regional case studies in Switzerland (Glanzer and Muhar 2006, p. 1).

Regional manager Michael Payer emphasises that the knowledge of local conditions that the project generated was of high quality, stressing that the meticulous preparation of students and teachers, and their ability to acquire accurate knowledge of regional structures, was something that the local population and the actors involved found to be particularly impressive (personal communication, May 15, 2012). Therefore, this carefully processed knowledge could serve as a good basis for successful scientific knowledge transfer (Fig. 3.3).

3.2.7.3 Integration

Successful integration was an important success factor for scientific knowledge transfer in “Life 2014”. Integrating the local population and different important political actors guaranteed that the project would produce results that are really relevant for regional allies. From the beginning, regional actors could participate in

Fig. 3.3 Thematic workshop. *Source* Leben 2014 (*Photo*)

the selection of topics that the project should focus on scientifically. Only those topics that the whole group selected as being the most important became part of the research—and not topics that had been pre-selected by students and researchers from outside. The processes of involving regional actors in the topic selection can be labelled as an ideal element of integration. Here, the demands of practice for science-based solutions could actually be directed towards research and dealing with appropriate scientific questions.

Within the integration process, scientific results are directed towards concrete utilization as well. Not all scientifically possible solutions are relevant for practice, nor are they necessarily politically feasible. What is crucial for integration is that there should be a selection from different scientific alternatives that are possible in principle. The scenarios developed by the students represent this kind of selection of scientific results in regard to practical aspects. Based on the most current information, the researchers developed different alternatives for every polarity field. Only the alternatives that won the most votes from regional actors in the course of a workshop were selected for further research activity.

In the polarity field "single and together", mentioned above, the researchers did not develop unrealistic concepts for a complete fusion of the nine communities into one. From a scientific point of view, such a central community could be one

Fig. 3.4 Theatre play from students during final meeting. *Source* Leben 2014 (*Photo*)

possible alternative that might suggest the highest possible level of cooperation. But it seemed to be more realistic and politically feasible to further develop the other scenario "Hand in hand in one regional association". As selected during the integration process by the regional actors, the project only developed concrete solutions to realize this scenario. The further development of the regional association has to be seen as result of successful integration activities.

To have a greater impact scientific results have to be intermediated in regard to different target groups by using different media and forms. The researchers in "Life 2014" communicated the project content in many forms. One very important fact was that students and researchers were present in the region. They worked in the Oberpinzgau region for 4 weeks and lived with local families. This, and the broad knowledge about the region that the students had, led to good cooperation and trust between regional actors and researchers. During the research process a trustful atmosphere between students, scientists, and regional actors emerged.

Special factors influencing target-group oriented intermediation were the naiveté and freshness of students, and the interest of local actors in interacting with these external young people in ways different from those in their daily conflicts in the region. Students' interaction with the regional population established a greater acceptance for the project and its results in the region. A key example for successful target-group oriented intermediation is the students' performance of a play about the polarity field "single and together" polarity field during the presentation of the project results (Fig. 3.4). They visited local farmers to find suppliers to organise a dinner consisting of organic food products. In this way, regional actors were made aware of some of the possibilities that could be further developed in the region.

Another factor of good target-group oriented intermediation is that all concrete projects were described in detail in the research report and that everything needed for successful implementation was considered (Michael Payer, regional manager, personal communication, May 15, 2012). The connection between theoretical and practical approaches was very well developed in the description of future regional development projects.

3.2.7.4 Utilization

The supply of utilization products is necessary for successful scientific knowledge transfer. At this point, scientists cannot make any changes, as the use of these products is in the hands of practical actors. In "Life 2014" students submitted different reports, and their suggestions for concrete regional development projects, in particular, to the regional actors. Regional actors and project partners agreed that, from then on, the choice between using the transdisciplinary projects' results, or developing them further, would be the responsibility solely of the region (Vilsmaier 2006, p. 278). The project represented very typical scientific knowledge transfer processes between research, integration, and utilization.

Utilization products represented criteria like appropriate solutions to problems, contribution to democracy, and contribution to the rule of law: They were created in a participatory process involving many regional actors (democracy), and were aimed at meeting societal goals, like sustainability. They could be realised within the confines of existing juridical norms and institutional background (for instance, realisation of regional cooperation in the context of the regional association that is strengthened and legitimised by the "Salzburger Raumordnungsgesetz", the Regional Planning Act of the state Salzburg). Utilization products also reflected appropriate solutions to problems: They produced solutions that could be applied to existing problems. "Life 2014" also led to the generation of scientific knowledge products: publications in scientific journals, contributions to scientific conferences, and many others. By means of these, scientists involved could connect to the scientific community of their respective fields and contribute to safeguarding the scientific quality of the project.

3.2.8 *Knowledge Transfer Success Factor List*

The following figure (Fig. 3.5) shows that all the main criteria for RIU were fulfilled in this process. If a particular criterion is of special relevance, it is marked "+".

Hand in hand - Life 2014 in the Pinzgau region
Areas of activity for knowledge transfer (RIU Model)

1	Research	
	Assessing current scientific information	✓
	Compliance to the procedures of good scientific practice	✓
	Cooperation with other scientific institutions and projects	✓
	Independent meaningfulness of scientific findings	✓

2	Integration	
	Orientation toward public goals (like sustainability, …)	✓
	Relevance in regard to political processes	✓ +
	Relevance in regard to allie (internal allies)	✓ +
	Target-group oriented intermediation	✓ +

3	Utilization	
	Contribution to democracy	✓
	Contribution to the rule of law (constitutional legality)	✓
	Contribution to "good governance"	✓ +
	Appropriate solutions to problems	✓
	Participation in the scientific discourse	✓

Fig. 3.5 Success factor list for the Hand in hand—Life 2014 in the Pinzgau region research project

3.3 Ski Tourism Within Climate Change—The STRATEGE Project

3.3.1 Project Fact Sheet

- **Project**

  STRATEGE – Climate change in winter sport

- **Type of project**

  Research project

- **Participating research institutions**

  - Institute of Landscape Development, Recreation and Conservation Planning of the University of Natural Resources and Applied Life Sciences, Vienna (*Institut für Landschaftsentwicklung, Erholungs- und Naturschutzplanung (ILEN) der Universität für Bodenkultur Wien*)
  - Austrian Institute of Spatial Planning (*Österreichisches Institut für Raumentwicklung*)
  - Manova GmbH (market research institute)

- **Location**

  Schladming, State of Steiermark

- **Central question**

  - What are the effects of climate change on the Schladming winter sport region?

- **Goals**

  - Development of scenarios based on climate models, regional analyses, and market research, to be discussed with regional stakeholders
  - Development of measures and alternatives to counteract potential negative effects of climate change in the region

- **Timeline**

  2005-2007

- **Financing**

  - proVISION – provision for nature and society, funded by the Austrian Federal Ministry of Science and Research (*Bundesministerium für Wissenschaft und Forschung*, BMWF)

3.3.2 *Winter Sports and Climate Change: The STRATEGE Project at Schladming*

Austria is a skiing country. This sport is highly popular and provides good business and employment opportunities within the mountain regions. However, in 2000 the public was alerted to the fact that the basis of ski tourism was in danger. Experts claimed that increasing temperatures due to climate change would melt the snow in winter. Even a slight increase in the average temperature would cause severe problems for the quality of snow at lower and middle altitudes in the mountains. Snow cannon technology, which had worked until then, would no longer be effective. Adaptation to higher temperatures would be technically difficult and expensive (Fig. 3.6).

In addition, a snowy trail running down an otherwise green slope would not meet the expectations of tourists longing for the magic of sparkling-white mountains. The most important regional newspaper, "Salzburger Nachrichten", reported a study by experts of the Organisation of Economic Development (OECD), which forecast a bleak future for ski tourism in the Alps in the Salzburg area (Kaindl 2006). The frequency and intensity of snowfall would be halved in areas over 1000 m above sea level.

The disaster warnings increased, stressing that a general warming of the climate would destroy the very basis of ski tourism. Tourists and the tourist industry were nervous about the future. Whereas the tourist industry and the owners of cable cars tried to downplay the threat, environmental groups mobilised the media and the

Fig. 3.6 Winter sports in Schladming. *Source* www.lanxx.at, Gernot Langs (*Photo*)

public, and the fear grew that the silent white winter paradise and the skiing might come to an end. This conflicting public discourse provided the impulse for the STRATEGE project. The project seeks to clarify scientifically how big the threat to ski tourism really is.

3.3.3 Content of the STRATEGE Project

The STRATEGE project focused on the specific impact of climate change on ski tourism within the mountain area of Schladming (Fig. 3.7). The interdisciplinary project looked at the ecological impact of temperature on snow and skiing conditions and at the demands and habits of tourists. Based on this information, the project sought to design innovative, alternative solutions for tourism in winter. The whole project was planned so as to be conducted in an interdisciplinary manner.

The supply and quality of snow are the key to skiing technology. Its dependency on temperature and moisture is highly complex. The first goal of the project was to shed light onto the complex relation between the local climatic conditions and the supply and quality of the snow for skiing. The second goal was to design alternative tourist uses for the alpine area, in case the snow were to diminish in the future.

The media discourse provided simple and alarmist answers to these two questions. The snow would melt due to climate change and winter tourism would face its end. The tourist development in Schladming, with more than 100 years of success, would fade away as a costly result of climate change.

Fig. 3.7 The Schladming glacier. *Source* Dachstein (*Photo*)

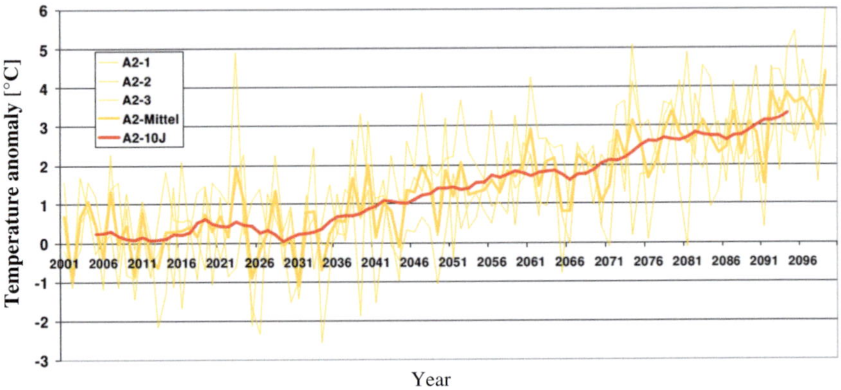

Fig. 3.8 Temperature scenario of the project. *Source* Pröbstl (2007, p. 3) (*Data* Formayer, Institute of Meteorology, University of Natural Resources and Life Sciences, Vienna)

In contrast to such general judgment, STRATEGE predicted a very specific scenario for the temperature trends in the area (Fig. 3.8). Derived from scenarios of climate change and the specific conditions of the alpine landscape, the specific temperature changes within the next 100 years differ significantly from a simple increase of temperature. During the first 30 years, from 2001 to 2030, not much will change. The average temperature will remain within a small range of increment of less than 1 °C. After 2030, there may be a linear increase in the average temperature, up by 3 °C toward the end of 2100 (Pröbstl 2007, p. 3). These would gradually worsen the conditions for snow cannons.

Local analysis further contradicted the existing assumptions. Climate change experts assumed that the climatic conditions for snow are altitude dependent linearly. Higher areas were considered safer, whereas lower areas would be in danger of losing the snow needed for the ski trails. STRATEGE proved that the specific physical type of the local countryside is more significant than the altitude alone. Depending on the exposition, vegetation, specific winds, morphological conditions and specific shelter provided by mountains, the cover and quality of the snow differ greatly (Pröbstl and Prutsch 2008a, p. 35). The general model of climate change had to be adapted to the local conditions in the mountain area. Additional, local measurements are indispensable in order to comprehend the specific snow conditions. The periods during which snow cannons can work properly depend on specific weather conditions that must be known in order to judge the potential for this technology correctly. Figure 3.9 shows the number of hours during which weather conditions favour the use of snow cannons (Pröbstl and Prutsch 2008a, p. 35).

The scientific results show that the weather conditions for preparing trails using snow cannons are better in the Schladming valley than at a higher altitude. This astonishing result is especially important during the months of December and January because of the peak demand for ski resorts at that time. In contrast, the conditions at the famous sky mountain, "Planai", at over 1300 m, are critical. In the future they will worsen, except in November (Pröbstl 2007, p. 4).

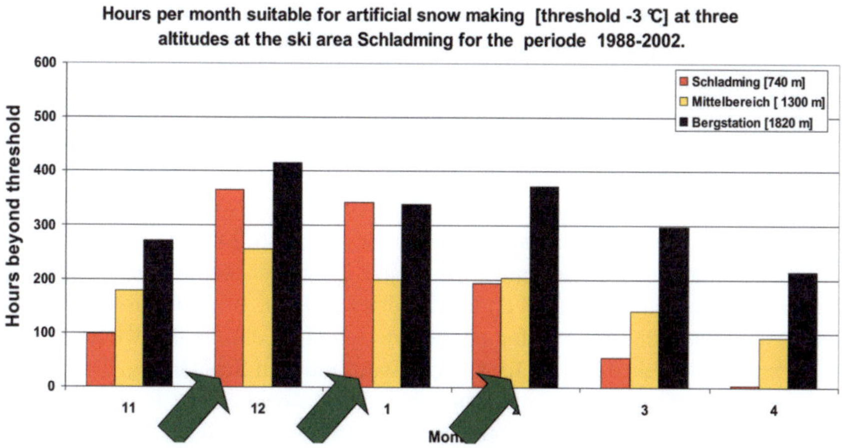

Fig. 3.9 Options for artificial snowmaking. *Source* Pröbstl (2007, p. 4) (*Data* Formayer, Institute of Meteorology, University of Natural Resources and Life Sciences, Vienna)

STRATEGE shows clearly that the conditions for winter sports will change after 2030. The traditional ski trails will have severe problems with snow, and snow cannons will no longer be a viable solution, both economically and in terms of efficacy. The future threat is real and scientifically proven, but the good news is that there are two decades to develop alternatives for winter sport.

What would a promising path for the development of winter tourism in the future look like?

STRATEGE analysed tourist demand in order to provide a scientific basis for the design of future alternatives for tourism. A survey showed that the main factors motivating the tourists to choose the Schladming resorts for winter holidays are "Guaranteed excellent snow on trails" and "Impressive white winter". Additional facilities like "Shopping", "Apres-Ski" or activities without snow are a minor attraction for the tourists (Pröbstl 2007, pp. 7–8). A strategy for securing tourism in the alpine area has to be built on such realistic information about the change in supply and demand. STRATEGE came up with three main priorities for the future of tourism (Pröbstl 2007, p. 8):

- Keeping up the potential of the winter tourism
- Looking for alternatives to winter sports
- Developing alternatives well in advance

Maintaining winter tourism requires innovative technical improvements in the trails and snow cannons. For example, making the trails more even allows for a thinner snow cover and saves a huge amount of artificially produced snow (Pröbstl 2007, p. 9). This solution follows an adaptation strategy for climate change. Another option is to transfer the trails to new areas where the snow conditions are better. These could be shadowy hillsides at much higher altitudes. Such an adaptation would be costly and carry ecological and economical risks. New trail areas disrupt the fragile ecosystem of the alpine zone. The cost of innovative snow cannons might be substantively higher than for the existing technology. Some of the predicted scenarios may not be accurate, and the trails may lack enough snow and turn green. Finally, technical adaptation would hinder other innovative alternatives and deepen the risk involved in engaging solely in ski tourism in areas where significant warming can be expected after 2030. The project concluded that technical adaption is fine for the near future and the period of transition but it is not considered to be a sustainable strategy for the region of Schladming (Pröbstl 2007, p. 9).

For the middle and distant future the project recommended an intensification of the search for alternative tourist options (Pröbstl 2007, p. 10). The survey revealed that there was little demand for alternative activities during the winter season because tourists looked mainly for skiing opportunities in Schladming. The project designed alternatives that covered the whole year, avoiding the focus on winter only. The offer for the tourists should combine activities for all seasons, e.g., every ski tourist could also get a voucher for a vacation in Schladming in the summer. To make this offer attractive the options for summer tourism had to be improved substantially.

A summary of STRATEGE provides scientific knowledge about the impact of climate change in the specific region of Schladming and about future options (Pröbstl 2007, p. 10):

- The scientific scenario shows for the first time exactly where the problems with snow and snow cannons will occur in the future, when this will happen and how strong the effects will be.
- Furthermore, the scientists show alternatives for the development of tourism in the region of Schladming in order to meet the challenges of climate change that will be relevant after 2030.

The project underscores the necessity for a climate change strategy for tourism in the alpine area, and it also shows that there is a time span of about 30 years to work on this transition. The scientific results draw a picture of the options for the future that is quite different from the panic scenario communicated by the media (Pröbstl and Prutsch 2008a, p. 40).

3.3.4 Main Impact: Science-Based Options for Regional Adaptation to Climate Change Instead of Panic

What is the outstanding effect of the STRATEGE project that would justify presenting it as success story in a book on science making the world go round? Climate change is a severe but unspecific threat. The high uncertainty of disastrous consequences tends to create anxiety in people and the media. The threat of an overwhelming catastrophe mobilises general fear but leaves people uninformed as to what to do precisely in their own lives and businesses. This was true for the climate change debate too. In 2003 the World Tourism Organization announced that the minimum altitude at which there would be enough snow for skiing would move up 150 m for each degree Celsius in warming in the future decades (World Tourism Organization 2003, p. 33). In Switzerland, famous for its skiing, it would only be possible at over 1600 m above sea level in 2030 (World Tourism Organization 2003, p. 47). Such climate-change driven developments are of specific relevance for Austria because tourism is one of its key industries. The economic consequences for profits, taxes and labour in alpine areas would be severe. Whereas the positive economic balance of the 2011/12 winter season set a new record,[3] such positive results will no longer be possible as the climate warms up in the future. Even if the tourism companies were persuaded by this negative prognosis, they could still not acquire information on what to do to ensure their businesses survive in the future. The message regarding the threat is repeated, but it provides not a hint as to what specific actions are needed to meet future challenges. Local cable cars and ski businesses need very specific information on the local conditions for snow and skiing in the future. STRATEGE chose answering these questions of high practical relevance as one of the main objectives of the scientific investigation. The research focusses on specifying the national climate models and the data on the regional and local conditions (Pröbstl and Prutsch 2008a, p. 35).

The media reports on the catastrophic loss of snow and trail quality in the Alps that has already caused a change in the demand by winter tourists. Driven by worry about not finding optimal skiing conditions, some tourists dropped formerly highly-rated tourist areas like Schladming as an option for their winter holidays (Pröbstl and Prutsch 2008a, p. 35). Accordingly, at the beginning of the STRATEGE project researchers and the critical public were convinced that the tourist industry would play down the negative consequences of climate change in order not to worry potential guests (Pröbstl and Prutsch 2008a, p. 37). The fear was that the tourist industry would avoid alarming and negative reports, even if they were based on sound science. The goal of keeping the tourists coming was dominant for the industry (Viktoria Haberl, Professional Association of Austrian Cable Cars, personal communication, July 14, 2011).

This behaviour was changed by the project. In the beginning, the industry expected from the project another round of critical messages announcing that the Alpine skiing would come to an end soon. However, in the course of the project, the specific information convinced the tourist experts that there was some value in the local future scenarios for snow and skiing conditions. They started listening to the scientific results and considering them as a basis for their own ideas. The former distance between industry and researchers vanished and there was even the beginning of an active collaboration toward the end of the project.

On 14 November 2007 a joint press conference (Fig. 3.10) was organised, the participants of which included the Chamber for Economy Austria, Professional Association of Austrian Cable Cars and Prof. Pröbstl-Haider, leader of the scientific

---

[3]According to "Statistik Austria", www.statistik.at/web_de/presse/064416, accessed on 19 Feb 2016.

Fig. 3.10 Joint press conference. *Source* Professional Association of Austrian Cable Cars 2007 (*Photo*)

project. The opening message makes clear why the cooperation came into being. The speaker of the Professional Association of Austrian Cable Cars stressed that the main results of the scientific project were that there would be no change in temperature within the Schladming ski area until 2030 and that all trails were safe and would provide the area's usual excellent skiing conditions. The facts presented at the beginning of the press release explain why the cable car industry overcame its scepticism in this case: "The most important results: No significant temperature changes are expected for the Schladming region until 2030, and the snowmaking possibilities do not depend on the altitude of the ski areas. Depending on the specific landscape, low altitudes may provide safe areas for effective snow cannons even under conditions of increasing temperature" (Professional Association of Austrian Cable Cars 2007, p. 3, own translation). The representative of the association stressed that the scientific project's contribution to the understanding of the consequences of climate change for the skiing area was highly appreciated: "Science prepared the way for a new evaluation of the problem based on facts". The specific information was understandable for the local actors. They felt that their experience had been recognised by the project and that there had been a serious effort in a joint search for solutions" (Viktoria Haberl, Professional Association of Austrian Cable Cars, personal communication, July 14, 2011, own translation).

3.3.5 *Factors for the Success of Knowledge Transfer*

STRATEGE was able to activate the following major factors to produce a successful transfer of scientific knowledge to political practice.

3.3.5.1 Transfer Through External Allies

STRATEGE managed a close cooperation with the central actor, the Professional Association of Austrian Cable Cars. This lobby association did not take part in the research process. The decisive role of the Professional Association of Austrian Cable Cars was to get other actors to accept the results of the scientific research and to make use of the scientific findings for their own activities. We call such an actor, who does not take part in the research but forces other actors to accept scientific findings, an "external ally".

The leader of STRATEGE, Prof. Pröbstl-Haider, stressed that they planned to win the Professional Association of Austrian Cable Cars as an ally for the project from the start. But in the beginning, the Association played down the negative consequences of climate change in order not to wake up fears in the winter tourists. They were not to become doubtful about the excellent snow and trail quality, and then choose another area as a ski resort. Nevertheless, the project managed to contact the Association and to inform them about the central findings. Once informed, the lobby organisation quickly realised that some of the scientific results about the specific snow conditions in the future could be quite useful for the interests of the Austrian cable car industry. STRATEGE predicted that the excellent snow and trail conditions would remain unchanged during the following 30 years. This was big news for all actors who made their living from winter tourism and who wanted to keep their cable cars businesses running profitably at least into the near future. A period of 30 years is a short time in relation to the process of climate change, but a long time span for business, approximately one generation of owners. Seen from the perspective of commercial interests, this scientific fact means that the problem can be postponed to the next generation of entrepeneurs. Such an option comes close to the message "problem solved". It was no wonder that once that the Professional Association of Austrian Cable Cars lobby group had understood the potential consequences of the scientific findings, it reversed its position and became an ally for the project.

In contrast to its sceptical position at the beginning, the Association started to cooperate and supported strongly the spectacular press conference in 2007 on skiing and climate change in Schladming. The key message was "Cable cars are willing to meet the chances of climate change" (www.kleinezeitung.at/steiermark/liezen/632801/index.do). This event signalled a change in the entire climate change strategy of the Association. From neglecting and downplaying the threat, the arguments turned toward defining climate change as a challenge that the cable car branch would meet actively by means of innovative technologies in the future.

The Association of Cable Cars Austria has a considerable influence on individual cable car enterprises, the majority of which are members of the association. The new strategy of the Association implies that it tried hard to inform the members about the scientific results of STRATEGE. Furthermore, the Association convinced many members to follow the new strategy and to switch from ignoring and downplaying climate change to adopting innovative adaptation technologies. The Association fought hard against the sceptical resistance of the members. It put some pressure on the members to take the scientific findings of STRATEGE seriously and to accept them as an opportunity. In this sense the Association became a strong external ally for the transfer of scientific knowledge of STRATEGE.

Professional Association of Austrian Cable Cars became an ally for STRATEGE not because it was driven by a love of science or strongly persuasive scientific arguments. The Association supported STRATEGE because of the usefulness of the scientific findings for supporting the interests of the Association. The RIU model argues that the fact that this cooperation was interest-driven does not in any way mean that the resulting transfer of information has no value. The scientific facts about the snow scenarios remain scientifically valid whether they favour the interests of the cable car companies or not. The scientific findings were recognised and had a practical impact. This is successful knowledge transfer from science to practice. The success is not diminished by an external ally's using its power, in pursuit of its own interests, to promote the transfer to individual cable car businesses.

3.3.5.2 Research

In our RIU model of successful scientific knowledge transfer, the research underlying the knowledge transfer must be of excellent quality, and it must cover the current scientific state of the art.

STRATEGE applied multiple scientific methods from climatology, economics, regional development sciences and tourism science. The highly relevant climatic data of STRATEGE were fully based on state-of-the-art climate research (according to W. Lexer, Environment Agency Austria, personal communication, April 27, 2012). STRATEGE adapted a new methodology in tourism research from Australian scientists. The innovative approach, called "TOMM" (Tourist-Optimisation-Management-Model) had never been tested in Europe before (Prof. Pröbstl-Haider, University of Natural Resources and Life Sciences Vienna, personal communication, July 12, 2010). All methods are well documented in the report on

STRATEGE (Pröbstl and Prutsch 2008b, pp. 11 ff.). The scientific discourse on congresses and in publications is an additional indicator for good science.

The results of STRATEGE were presented in numerous national and international seminars (Pröbstl and Prutsch 2008b, p. 76 ff.). Key findings were published in journals, e.g., the survey on winter tourists was published in the international journal "Tourism Review" (Unbehaun et al. 2008); a strategy for adaption was published in the journal "Naturschutz und Landschaftsplanung" (Landauer and Pröbstl 2008).

STRATEGE was part of the scientific programme "proVision", which was launched by the Austrian Ministry of Education and Science. The proposal passed successfully the scientific evaluation and met the research milestones that control scientific standards and progress.

Furthermore, STRATEGE cooperated with scientific partners, one of which, the research institution "Manova" is well recognised in market research. Another partner, the Austrian Institute for Regional Research (ÖIR) has a long-standing reputation for scientific research. It provided theoretical and methodological knowledge on issues of regional development of the area of Schladming.

The good scientific reputation of STRATEGE fostered cooperation with the actors in the field. They could trust that the findings of STRATEGE were scientifically sound and could stand up well to critique by other scientists and experts.

3.3.5.3 Integration

STRATEGE linked science to practice well due to the strong orientation of the research toward practical needs. The project made the right assumptions about the potential of the research and the needs of actors in the field, and managed to increase their compatibility. This integration process was successful in its selection of specific research questions and specific results which met needs in practice.

STRATEGE took its general orientation from the public goal of sustainable development. Climate change threatens the development of the Schladming tourist area and STRATEGE aimed to determine the true dimension of the future threat of increasing temperature and to design an innovative, sustainable way for the region to survive in the future. The goal of sustainable development, of great public importance, drove STRATEGE to research questions of public interest, fostering the interest of those in political practice as well.

STRATEGE focuses on problems specific to a ski area. The research became specific in questions of local climate, local landscape, planning of trails and the technology of snow cannons. It looked directly at the problems that winter sports and tourism had dealt with for the previous decade.

In addition, STRATEGE approached the specific positions of actors in the area of Schladming. The owner and managers of small businesses of cable cars, trails and hotels were able to recognise their specific position, problems and interests within the selected research questions and results of STRATEGE. The selection had its unavoidable price. The integration process eliminated from the project, very early, some very interesting research questions and ignored some of the demands of actors in the field. But what was selected in the end fulfilled the expectations of meeting the demands of specific actors and the potential of science simultaneously. In particular, the chosen focus on the Association of Cable Cars Austria narrowed the research question down and won support by this powerful external ally for specific results only. Nevertheless these results selected in the integration process had the chance of being transferred into practice due to their scientific content.

STRATEGE managed to present the results in an accessible language and in the media, which can reach out to the target groups. Reports and personal consulting reached the specific community of tourism experts. The experts had no need to learn the scientific language first. The scientific message was presented directly in their language and using specific technical terms that corresponded to their view of the world. Additionally, the project aimed to discuss results frequently within the local context of Schladming. From the beginning, the integration selected specific issues with the potential of being very relevant for the local community.

3.3.5.4 Utilization

STRATEGE needs to produce specific information that can be used in practice, in order to transfer scientific knowledge to the practice of ski tourism. Such information is packed in "utilization products" like reports, written advice or films. An important aspect of the transfer is that as soon as the information product is given to an actor, scientists cannot influence the information process anymore. The actor is autonomous in gleaning the information as he sees fit.

STRATEGE presented the scenarios about the future of skiing in the Schladming area in different reports, for specific target groups and actors. The reports reached the actors directly and were utilized by them.

The content described innovative and sound local solutions relevant for each specific actor. They had been designed in a participatory process, integrating regional actors and the public in a democratic manner.

In addition, STRATEGE took the economic and legal systems into account. The reports presented solutions in such a way so as to adapt to the applicable legal constraints. Furthermore, the reports informed about the economic rationale of the solutions. They included direct links to the small business firms and to the economy of Schladming. Finally, STRATEGE published successfully in scientific journals, linking the project to the scientific discourse that contributed to the scientific quality of the project.

3.3.6 Knowledge Transfer Success Factor List

The following figure (Fig. 3.11) shows that all the main criteria for RIU were fulfilled in this process. If a particular criterion is of special relevance, it is marked "+".

STRATEGE
Areas of activity for knowledge transfer (RIU Model)

1 Research

- Assessing current scientific information ☑ +
- Compliance to the procedures of good scientific practice ☑
- Cooperation with other scientific institutions and projects ☑
- Independent meaningfulness of scientific findings ☑

2 Integration

- Orientation toward public goals (like sustainability, …) ☑
- Relevance in regard to political processes ☑ +
- Relevance in regard to allies (external allies) ☑ +
- Target-group oriented intermediation ☑ +

3 Utilization

- Contribution to democracy ☑
- Contribution to the rule of law (constitutional legality) ☑
- Contribution to "good governance" ☑
- Appropriate solutions to problems ☑
- Participation in the scientific discourse ☑

Fig. 3.11 Success factor list for the STRATEGE research project

3.4 Indicators for Human Impact on the Biosphere and Sustainable Development

3.4.1 Project Fact Sheet

Project

The colonization of nature – Indicators for sustainable land use

Type of project

Research projects

Participating research institutions

- *Indicators for sustainable land use:*
- The Institute of Social Ecology of the Faculty for Interdisciplinary Research and Further Training (IFF), University Klagenfurt (*Institut für Interdisziplinäre Forschung und Fortbildung, Abteilung Soziale Ökologie (IFF) der Universität Klagenfurt*)
- Austrian Ecology Institute, Vienna

The colonization of nature

- Sustainability Studies Vienna (*Forschungsinitiative nachhaltige Entwicklung (FINE) Wien*)
- The Institute of Social Ecology of the Faculty for Interdisciplinary Research and Further Training (*Institut für Interdisziplinäre Forschung und Fortbildung, Abteilung Soziale Ökologie,* IFF),
- University of Klagenfurt (*Institut für Interdisziplinäre Forschung und Fortbildung, Abteilung Soziale Ökologie (IFF) der Universität Klagenfurt*)

Location

Vienna

Central question

- How can sustainable development be operationalized by means of indicators?

Goals

- Development of polluter-oriented indicators for the measurement of physical streams between nature and economic sectors
- Empirical description oft he colonization of nature in Austria with the help of the indicator, human appropriation of net primary production (NPP)

Timeline

Polluter-oriented land use indicators: 1989-1991

The colonization of nature – Indicators for sustainable land use: 1997-1999

Financing

- Polluter-oriented land use indicators:

Austrian Federal Ministry of the Environment, Youth and Family (*Österreichisches Bundesministerium für Umwelt, Jugend und Familie*)

- The colonization of nature – Indicators for sustainable land use:

Cultivated Landscape Research (Kulturlandschaftsforschung, KLF), sponsored by the then Austrian Federal Ministry of Education, Science and Culture (*Österreichisches Bundesministerium für Bildung, Wissenschaft und Kultur*)

3.4.2 *Information About Human Impact on the Environment*

Everybody is familiar with this situation: You feel poorly and you measure your body temperature for an indication of whether you are seriously ill. A high temperature indicates that a virus or bacteria is causing problems for your body. Acquiring a clinical thermometer is easy and the measurement is simple but the information that something in your body is wrong is quite valuable. An alarm is sounded even if the causes of your illness are unclear.

Such measured data, which afford a quick insight into the state of a system, are called indicators. "Indicators are statistical data that are measurable and represent important parts of a subject of research" (Rogall 2008, p. 213, own translation). Indicators are frequently applied (Turnhout et al. 2007): The mean temperature is used as an indicator for climate change. Measurements of the concentration of specific chemical elements in the air are used as indicators for the quality of the air. This works fine as an alerting system even if the source of pollutant emissions remains unrevealed. Indicators provide information on the state and development of systems; they simplify and focus complex processes (Haberl et al. 2001, p. 9). Indicators work for early warning systems used by environmental protection and nature conservation (Garrelts and Krott 2002, p. 114). They make problems visible, stressing the need for political action.

> The idea behind indicators is to quantify a phenomenon which is too complex to be measured directly. Furthermore, indicators should communicate problems for sustainable development. They show whether the development of ecology meets specific standards over time (Haberl et al. 2001, p.10).

Indicators are frequently used in economics. The gross domestic product (GDP) is a simple number that characterises the strength of a complex entity like the entire national economy. The indicator cannot provide information as to which processes caused a specific level of GDP. Another important economic indicator is the unemployment rate. The unemployment rate (expressed as a percentage) says something about an economy's state and its level of employment, but the reasons for the number of unemployed people cannot be registered by this indicator.

At the start of the project in the 1980s the indicators of development in economic terms were well recognised, but nothing was known about the ecological consequences of economic development. The success of environmental politics and the increased influence of environmentalists started a professional and public discourse on the need for quantitative indicators for the ecological impact of human activity. They stated that the calculation used to determine the GDP was wrong without including the proper data on the negative impact on the environment (Rogall 2008, p. 217; Spangenberg 1996).

Major omissions in the calculation of the GDP mean that this indicator is worthless for judging the sustainable development of a country correctly. For example, the life span of products is not integrated into the calculations; two lawn mowers with a life span of 5 years each will result in the calculation yielding a much higher contribution to the GDP that one lawn mower of better quality and a life span of 10 years. In addition, the economic indicator does not provide information on the amount of natural resources used for production.

The ecological blindness of the calculation of the GDP triggered the professional and political issue of the integration of the ecological dimensions into it (Rogall 2008, p. 224). The connections of economic development to the ecological dimension should be calculated properly.

In the 1980s this issue was also taken up by politics in Austria. Deficits in the description and measurement of the ecological impact of the Austrian economy were stressed and the way such impacts could be measured was discussed. The parliament discussed this issue and pushed politically to start a scientific project to clarify options for indicators of human impact on the biosphere and sustainable development.

3.4.3 Political Drivers for Measuring Human Impact

In 1987 the members of the Green Party in the Austrian Parliament put forward a motion for measuring the human impact of the economy. The Ministry of Environment was to contract out such a project to scientists. The Member of Parliament, Dr. Bruckmann of the Austrian Conservative Party, gave a critical statement on the problems of measuring the human impact on the environment: "The gross national product is considered by laymen often as a big cake of products and services. A bigger cake makes Austrian people happier. If the growth of the gross national product exceeds the data of other countries the media announces and celebrates the good news that Austria is on the right way toward surpassing other national economies. The scholars in economics know that the gross national product is not a sound indicator for the wellbeing of a nation. But they do nothing to criticise the gross national product and to replace this most important indicator by a better one" (Stenographic protocol of the session of the Austrian Parliament, March 22, 1988, own translation).

Bruckmann added a striking example of the misleading information provided by the gross national product: "We know that a serious car accident increases the gross national product. The car workshop gets work, in the same way that the hospital, and in the best cases the cemetery, gets an economic push". He made a comment on the ecological impacts too: "If the traffic through the mountain valley in Tyrol doubles and the emissions kill trees and the noise disturbs people and make them sick, the gross national product shows a strong plus but the damage to forests and people are ignored. Relying on the data of the gross national product draws a wrong picture of the development of Austria and is misleading the judgment of politics and the public about the right way to go forward" (Stenographic protocol of the session of the Austrian Parliament, March 22, 1988, own translation).

The prominent speaker of the Green Party Austria, Ms Freda Meissner-Blau, responded: "We have to stop this mess with statistics. We do not want to abolish the data of the gross national product, we are looking for the opposite. We want to add good data about the impact on the environment. These data are needed to discriminate between the good and the bad of economic development" (Stenographic protocol of the session of the Austrian Parliament, March 22, 1988, own translation).

The motion by the Green Party Austria got the support of other parties and was approved. The task was two-fold. On the one hand the Austrian Central Statistical Office had the mission of keeping the established system for calculating the gross national product, but to expand it by calculating the impact on the environment in a new and proper way. A new input-output model had to be developed to integrate the environmental aspects into the existing terms and system. On the other hand the Ministry of Environment was to contract out a research project on the "social costs and benefits" which are relevant for the environmental impact of the economy (Stenographic protocol of the session of the Austrian Parliament, March 22, 1988).

The political discourse and the Parliament expressed clearly the necessity to improve the system used to calculate the national product and the need for stronger research in this issue.

3.4.4 *Project in Causative Indicators and Colonisation of Nature*

The initiative for research into the innovative measurement of the gross national product produced many projects that we will look at because they were highly successful in transferring scientific results to politics. Prof. Haberl of the Institute of Social Ecology of the University of Klagenfurt, Austria, took part in many of the projects. The research started with a project on environmental indicators that focussed on causative factors (Fischer-Kowalski et al. 1991). It was financed by the Ministry of Environment Austria and was conducted together with the Austrian Institute for Ecology. The new indicators should show the impact of human activities on the environment. The only existing indicators up to this time were an indicator of the impact of different sectors and cars on the quality of the air, which was mainly a register of emissions. In respect of the national accounts mentioned before, the project was innovative in one way: It did not develop indicators to represent monetary flows, but instead for the representation of the physical flows between nature and economic sectors (Fischer-Kowalski et al. 1991, p. 1). For example, in the area of ecologic-economic system indicators, indicators were developed for material intensity (Fischer-Kowalski et al. 1991, p. 14): These are purportedly important, because "most environmental damage is caused by extraction, transport, processing and utilisation of materials" (Fischer-Kowalski et al. 1991, p. 14, own translation). Indicators for material intensity were then proposed as being important indicators for the further development of the national accounts. These relate to the total input, material wear, material effectiveness and packaging intensity (Fischer-Kowalski et al. 1991, p. 18).

Total input refers to the amount of material used (advance provisions from nature, like directly extracted raw materials, or advance provisions from the economic system, like energy sources or waste materials), and material wear to the balance between input and output, the sum of all waste resulting from the production process. Material effectiveness is the proportion of all product output in relation to the input. Packaging intensity represents the proportion of directly used packaging material in relation to the total product output (all definitions are simplified according to Fischer-Kowalski et al. 1991, p. 18). These indicators provide a new basis for expanding the calculation of the gross national product to include the impact on the environment (Fischer-Kowalski et al. 1991, p. 2).

Another project continued the design of indicators with a focus on sustainable development. The project "Colonisation of cultural landscape" is aimed at specifying the general term "sustainable development" by means of quantitative indicators (Haberl et al. 2001, p. 7). The project defined sustainability in terms of the flow of physical sources. "Sustainability requires that the flow of physical sources and energy between the socio-economic system and the natural system can be kept stable during a longer period of history" (Haberl et al. 2001, p. 7, own translation). The interaction between the socio-economic system and the natural system are the focus. The social and economic processes are defined as driving forces and the impact on nature as pressure on the environment.

One innovative element of the concept is the "socio-economic metabolism." (Haberl et al. 2001). This is the way in which societies organise their exchange of matter and energy with their natural environment. The other innovative element is the "colonisation of nature" (Haberl et al. 2001). This is the group of strategies employed to transform parts of the environment in order to render them more useful for societal needs. An example of colonisation would be building a road or transforming forests into agricultural land.

The focus is on pressure-indicators that should be relevant for the development of cultural landscapes. The changes in landscapes cannot be measured by the use of resources because the resources are modified within the landscape but they are not used and diminished. The results are new types of ecosystems within the landscape. The indicators should be sensitive to species, nutrition or water supply (Haberl et al. 2001, p. 19).

The changes are described as "human appropriation of net primary production (HANPP)" (Haberl et al. 2001, p. 24, 2007). The net primary production is the biomass, measured in energy units, which green plants produce in a given time period through photosynthesis. The primary production feeds energy into ecosystems used by the chain of nutrition.

The use of ecosystems by humans has consequences for the ecosystems. The overall production of biomass by green plants can diminish if housing development reduces the size of green areas. Harvesting of timber or crops means a loss of biomass that could go directly into the nutrition chain.

The human appropriation of net primary production is defined as the energy that is taken out of the ecosystem by human activities (Haberl et al. 2001, p. 24). The reference is potential vegetation that would otherwise grow without any interference by human activities.

The project shows that 90 % of the human appropriation of net primary production is caused by agriculture and forestry in Austria (Haberl et al. 2001, p. 24). For the first time the human appropriation of net primary production has been calculated for the historical period from 1830 to 1995. In addition, the spatial distribution in Austria in the year 1990 is shown (Weisz et al. 1999, p. 38). The final report gives an overview of the historical development (Fig. 3.12).

Figure 3.12 shows that the ratio of the human appropriation of the primary production to the potential of the human appropriation of the primary production has not changed significantly in the last 150 years (Fischer-Kowalski et al. 1999, p. 128). The explanation given is that the major changes in agriculture in Austria had already happened before 1830 in Austria (Weisz et al. 1999, p. 54). The transformation of forests into agricultural land had been carried out much earlier, in the Middle Ages. Since the Second World War, the harvest of biomass from primary production (ANPPe) doubled, due to the innovation in agricultural technology

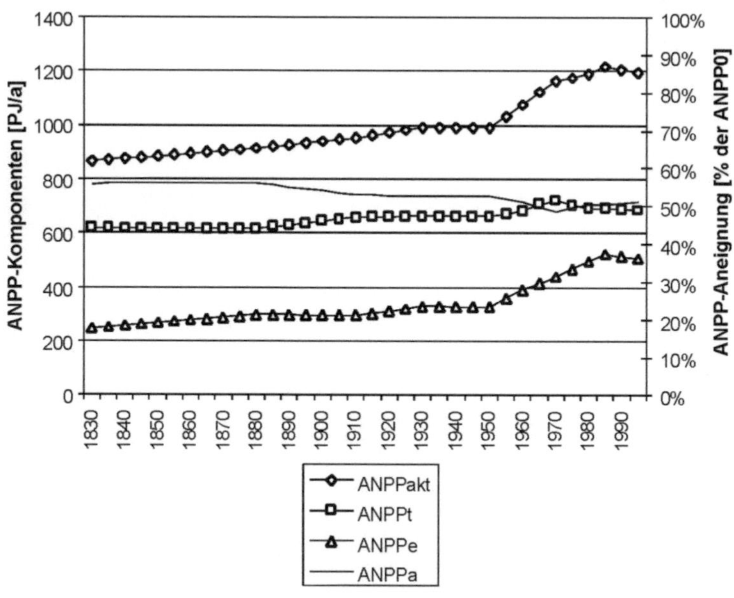

Fig. 3.12 Human appropriation of net primary production in Austria, 1830–1995. *Source* Fischer-Kowalski et al. (1999, p. 130)

and fertilisation (Weisz et al. 1999, p. 40). The productivity per area increased, indicating a growing socio-economic metabolism (Weisz et al. 1999, p. 40). The conclusion is that it is necessary to discriminate three aspects of the human appropriation of the primary production (NPP): the NPP of the actual vegetation, the amount of NPP harvested by humans and the differences in the NPP of the ecosystem in absence of human activities. The three indicators relate to different aspects of sustainability of the landscape (Weisz et al. 1999, p. 48).

These indicators are global "headline indicators" for the colonisation of terrestrial ecosystems (Weisz et al. 1999, p. 49). They show two types of areas: (1) highly productive areas which are not used much (e.g., forests) and (2) intensively used areas which have different levels of productivity. Within the intensively used areas little energy remains for keeping up the ecosystem itself. As judged by means of these indicators, the development in Austria in the last century is sustainable. No growing pressure on landscape was found (Weisz et al. 1999, p. 49).

The area for agricultural use had been shrinking until 1995, but the energy removed through harvesting was increasing. These intensive agricultural techniques are relevant for Austrian sustainability (Weisz et al. 1999, p. 49). Figure 3.13 shows the spatial distribution of the NPP in Austria in 1990.

The spatial distribution shows that the highest NPP harvesting is in intensive agricultural areas and the lowest in forested areas. Forests save the energy produced by biomass best (Haberl et al. 2001, p. 25).

The results of the project suggest that the calculation of the gross national product can be improved by the use of environmental indicators. This would be a

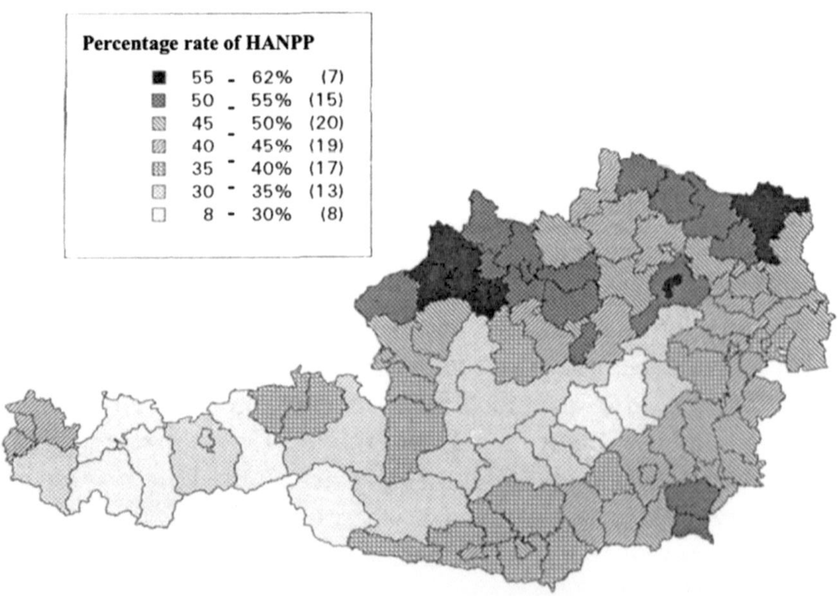

Fig. 3.13 Spatial distribution of the NPP in Austria in 1990. *Source* Haberl (1997b, p. 321)

major contribution of the statistics to the support of sustainable development in Austria. The goal is for the indicators to enable society to watch itself critically. Society and political actors can watch the impact of economic development on nature (Haberl et al. 2001, p. 48).

3.4.5 Impact on Improved National and International Public Statistics

The project was driven by the goal of improving the environmental indicators within the public statistics in Austria. It took a considerable amount of time before the knowledge about new ecological indicators made its way to application in public statistics.

In 1989 a department for environmental statistics was founded within the Austrian Central Statistical Office. This department cooperated closely with the envisaged project. It looked at the methods and indicators and adapted them for the public statistics (information from Ms Milota, Central Statistical Office Austria, personal communication, November 4, 2013). One application is the calculation of material flows in the Austrian economy (Milota and Petrovic 2012) (Fig. 3.14).

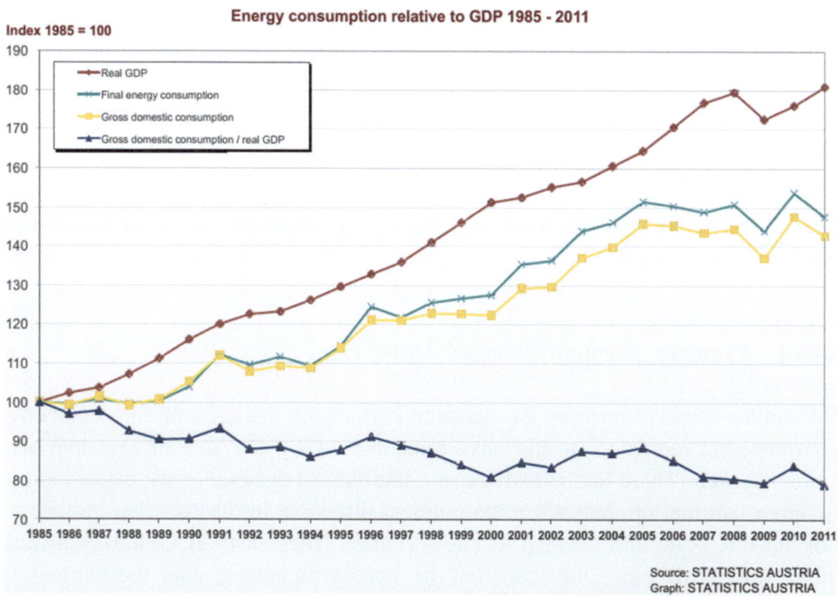

Fig. 3.14 Material flows in Austria 1995–2011. *Source* Federal Ministry of Agriculture, Forestry, Environment and Water Management (2013, p. 32)

The Austrian Federal Ministry of Agriculture, Forestry, Environment and Water Management also made use of the new approach in its reports on environmental accounting (BMLFUW 2010).

The department for environmental statistics still cooperates with the researchers of the project. A most recent product is the report on "Use of natural sources in Austria" (BMLFUW and BMWFJ 2011).

The indicators that were developed by the Austrian project were applied internationally. EUROSTAT, the agency of the European Union for statistics, integrated the indicators in its handbook of methods for calculating the flow of materials (European Commission 2001; Eurostat 2009). The scientifically designed innovative indicators became part of the statistical routine of international bureaucracies.

The scientific calculation of the human appropriation of the primary production was developed further and applied to a global analysis. A map of the global human appropriation of the primary production was published by the European Environmental Agency (European Environment Agency 2010, p. 131; see Fig. 3.15).

The leading scientist, Prof. Haberl, became a member of the Intergovernmental Panel on Climate Change (IPCC). Haberl was the lead author of the chapter "Agriculture, Forestry and Other Land Uses" of the Assessment Report until 2014. He evaluated the global research on the colonisation of nature by agriculture and forestry and summarised major findings in the report (Prof. Haberl, Institute of Social Ecology of the University of Klagenfurt, Austria, personal communication, May 14, 2012). It is important to evaluate the degree of scientific consensus on methods and results. This activity offers the chance for the indicators to be evaluated and accepted as having current validity in environmental statistics.

3.4.6 Factors for the Success of Knowledge Transfer

The transfer of knowledge from the project to political practice was fostered by specific success factors.

3.4.6.1 Transfer Through External Allies

Without the decisive push by the Austrian Parliament, the research into innovative environmental indicators would have remained within the area of excellent academic research. There had already been a tradition of critique of the calculation of the gross national product when the political discourse in the Austrian Parliament took up this issue and decided to effect change. The Austrian Central Statistical Office was to develop indicators for the impact on nature with the support of science. By putting administrative resources and money behind this whish, the

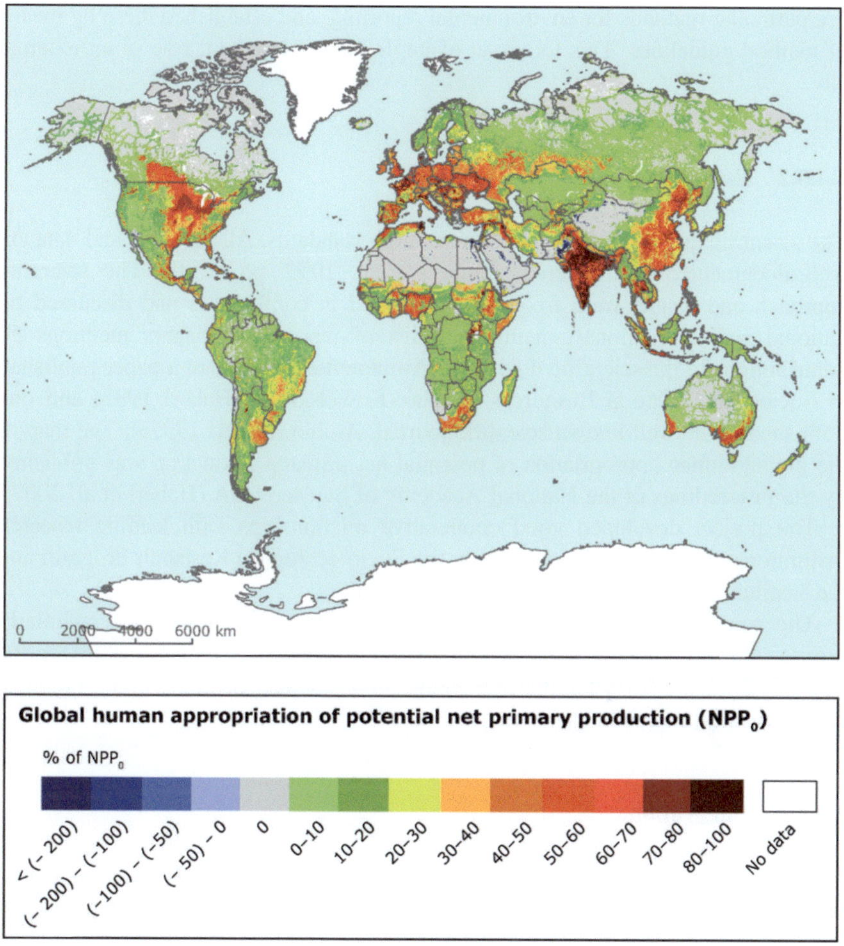

Fig. 3.15 Global human appropriation of potential net primary production. *Source* www.eea.europa.eu/data-and-maps/figures/global-human-appropriation-of-net

Parliament became a strong external ally for the transfer of scientific results in the form of relevant environmental statistics. The Parliament did not engage directly in the research or in designing statistics, therefore it played the role of an external ally perfectly. It is quite striking how effective the Parliament was. Such an involvement in specific political projects cannot be observed often but when it happens the Parliament can become a strong political actor.

As a follow up to the Austrian initiative, the European Union also became interested in such indicators. The European Union obligated the member states to use particular methods for environmental reporting, and established these by means of method guidelines. This too is an example of the important role of an external ally.

3.4.6.2 Research

The scientific quality of the project meets high standards. All methods and data are well documented in the report (Haberl et al. 1999, pp. 32 ff). The scientific approach and results were frequently presented in conferences and discussed by national and international scientists. A list of reports and papers mentions 50 contributions to the scientific discourse. Prominent examples are a paper published in Society and Natural Resources (Fischer-Kowalski and Haberl 1997) and one published in the well-known scientific journal, Ambio (Haberl 1997a). The map of the global human appropriation of potential net primary production was published by the Proceedings of the National Academy of Science USA (Haberl et al. 2007).

The project developed good cooperative relationships with leading research institutions. Partners in Germany were the Socio-ecological Research Program and the Institute of Ecological Economics in Berlin.

The scientific activities indicate that the project had an excellent scholarly profile. Our RIU model stresses that excellent science is indispensable for the successful transfer of scientific knowledge into practice. In this project, practical relevance did not conflict with research quality.

3.4.6.3 Integration

The scientists were strongly oriented toward the public goal of sustainable development. The whole career of Prof. Haberl is built on the task of promoting sustainable development through excellent science practice. His home institute in Austria is devoted to science that deals with problems of sustainable development. This idealistic orientation provided a general background for the researcher to take notice of the environmental issues brought up by the Green Party Austria and of the initiative of the Austrian Parliament to improve upon the calculation of the gross national product by introducing environmental indicators.

The initial momentum for the project came from the Green Party, within the Austrian Parliament. The Austrian Parliament became a strong external ally by directly prioritising the needs of research. The researchers had close contact to the Ministry of Environment, were notified early and designed a proposal that adhered to the highest standards. The positive orientation toward the political process and the direct focus on the needs of the Ministry of Environment allowed a good integration of the scientific potential with the political demands for specific topics and results.

The researchers communicated directly with the Austrian Central Statistical Office. The research was designed to meet exactly the needs of the Austrian Central Statistical Office in generating sound environmental statistics. The newly founded department for environmental statistics became a close internal ally who took the suggestions for indicators and adapted them to form a professional statistic method.

Finally, the researchers did not develop theories only to improve the statistics. They also integrated practical needs by basing their design on new, sound theoretical approaches to specific indicators that the statistical agency could understand well. The scientists produced a map of the global human appropriation of potential net primary production that was easily understandable for all statisticians around the globe. The use of an accessible language for a specific target group of international professionals made the international application of the results possible.

The researchers did not forget the national media as target group. They presented the results of the project successfully in the Austrian quality newspaper, der Standard (Der Standard "Eine neue Währung der Ökologie", February 27, 2007). By choosing the right style of media communication, the project made an important contribution to the ecological discourse in Austria. This experience supported the project in meeting the demands of political actors in green issues.

3.4.6.4 Utilization

The strong impact of the project on national and international environmental statistics was shown already. Two major success factors were the support by a democratic decision within the Austrian Parliament and the professional bureaucratic implementation of the mission. Such a direct political mandate is rare for research, but in this case it could lead to a good transfer for scientific knowledge.

Additionally, the researchers were well aware of the green discourse and linked their indicators to the green discourse within the public. They were supported on the improvement of instruments for measuring the gross national product by both the professional and public discourse. Improved indicators for national wellbeing and sustainability were strongly demanded by many environmental initiatives.

The indicators were a professional instrument to measure the complex metabolism of the socio-economic system and nature. They offer a sound and professional solution for a complex measurement problem.

Finally, the researchers took part in the scientific discourse that spread the method to other researchers on a global level. How much further they can go is still a challenge for the Austrian scientists.

3.4.7 Knowledge Transfer Success Factor List

The following figure (Fig. 3.16) shows that all the main criteria for RIU were fulfilled in this process. If a particular criterion is of special relevance, it is marked "+".

Indicators on the environment and sustainability
Areas of activity for knowledge transfer (RIU Model)

1 Research

Assessing current scientific information	✓	+
Compliance to the procedures of good scientific practice	✓	
Cooperation with other scientific institutions and projects	✓	
Independent meaningfulness of scientific findings	✓	

2 Integration

Orientation toward public goals (like sustainability, …)	✓	+
Relevance in regard to political processes	✓	+
Relevance in regard to allies (external allies)	✓	+
Target-group oriented intermediation	✓	

3 Utilization

Contribution to democracy	✓	
Contribution to the rule of law (constitutional legality)	✓	
Contribution to "good governance"	✓	
Appropriate solutions to problems	✓	
Participation in the scientific discourse	✓	+

Fig. 3.16 Success factor list for indicators on the environment and sustainability project

3.5 Options for Viennese Agriculture and Horticulture

3.5.1 Project Fact Sheet

Project

Options for the development of Viennese agriculture and horticulture

Type of project

Research project

Participating research institutions

- Ludwig Boltzmannn Institute for Organic Farming and Applied Ecology, Vienna (*Ludwig Boltzmann Institut für Biologischen Landbau und Angewandte Ökologie Wien*)
- Institute for Spatial Planning and Rural Development of the University of Natural Resources and Life Sciences (*BOKU*), Vienna (*Institut für Raumplanung und Ländliche Neuordnung der Universität für Bodenkultur* (*BOKU*) *Wien*)
- Institute for Interdisciplinary Research and Continuing Education, Department of Social Ecology (*IFF*) of the University of Klagenfurt (*Institut für Interdisziplinäre Forschung und Fortbildung, Abteilung Soziale Ökologie* (*IFF*) *der Universität Klagenfurt*)
- Institute for Finance Studies and Infrastructure Policy of the Technical University of Vienna (*Institut für Finanzwissenschaft und Infrastrukturpolitik der Technischen Universität Wien*)

Location

Vienna

Central question

What does the future role of Viennese agriculture look like?

Goals

- To design options for the development of Viennese agriculture
- Sustainable development of open areas in Vienna

Timeline

2000-2002

Financing

- Cultivated Landscape Research (*Kulturlandschaftsforschung*, KLF), sponsored by the then Austrian Federal Ministry of Education, Science and Culture (*Österreichisches Bundesministerium für Bildung, Wissenschaft und Kultur*)
- Municipal Department 22, Environmental Protection, of the City of Vienna (*Magistratsabteilung 22, Umweltschutz, der Stadt Wien*)

3.5.2 *Starting Point: Insufficient Knowledge About the Significance and the Lack of Strategy of Agriculture in Vienna*

When one thinks of agriculture and Vienna, one immediately thinks of the numerous Viennese wines, among which are many of very high quality. However, few know that, beyond wine, Viennese agriculture is of great significance for Vienna and Austria. Approximately 17 % of Vienna's territory is used in agriculture (Kromp et al. 2004, p. 21). Viticulture, tillage and market gardening (commercial, especially glasshouse, horticulture), parts of the areas for which are in the middle of the urban area, characterise agricultural usage in Vienna (Fig. 3.17).

In Vienna there are 700 hectares of vineyards alone (information from Dr. Seher, University of Natural Resources and Life Sciences, Vienna, personal communication, July 12, 2010). However, Viennese agriculture is threatened by changing conditions. These include, among others, falling prices for agricultural products because of increasing integration in the global market and problems particular to agriculture in the city, for example, traffic problems or land loss due to competing demands for use (Kromp et al. 2004, p. 21). For example, something typical of urban development is to provide land for construction projects (e.g., housing). In this way, and also in Vienna, arable land is supposed to serve as a reserve for future construction projects (information from Dr. Seher, University of Natural Resources and Life Sciences, Vienna, personal communication, July 12, 2010). Nevertheless,

Fig. 3.17 Viticulture in Vienna. *Source* LK Wien, Rausch-Schott (*Photo*)

among the Viennese areas used for agriculture, vineyard zones cannot be converted, since they have high-level protection status from the government.

The significance of Viennese agriculture due to its area (one sixth of the total area of Vienna!) was not reflected in the 1990s in a corresponding urban development that would consider agricultural lands commensurately. The origin for a stronger regard of agriculture within the city's developmental policy was that, due to its location in the city, it was subject to pressure to convert agricultural land to land for future urban development. On the other hand, however, agriculture was also a cause for repercussions in matters having to do with nature conservation. Therefore, the Municipal Department 22, Environmental Protection, was very interested in gathering information as to how it might deal with agriculture from a nature conservation perspective (information from Dr. Kromp, Institute Director, BioForschung Austria, personal communication, July 20, 2010).

In addition, conditions for agriculture in the 1990s had changed as a whole, due to the country's becoming part of the EU in 1995. Therefore the project's first concern was to raise the status quo of Viennese agriculture and its relevant framework conditions, especially against the background of the consequences that the future integration of the middle and eastern European countries into the EU could have (Maurer et al. 2002, p. 12).

At the same time, the notion of increasing the share of organic farming already existed in the political realm, not least because organic agricultural production could be rewarded with EU funding as a result of accession to the EU (information from Prof. Maurer, then at the Ludwig Boltzmannn Institute for Organic Farming, personal communication, July 22, 2010). Another determining factor was the referendum against genetic engineering in 1997, the results of which were adopted in 1998, and which rejected particular agricultural conceptions.

There were therefore many processes that caused the significance of agriculture in a major city like Vienna to be rethought, among other things. Some of the questions raised were those of what significance agriculture has in a major city in the first place, what conflicts of exploitation exist with regard to agricultural land, and how one could increase the share of environmentally sound agriculture and organic farming.

3.5.3 The Project: Options for Viennese Agriculture

The goal of the project, financed by the Austrian Ministry of Science and by the Municipal Department 22, Environmental Protection, of the City of Vienna, was to research the future role of agriculture in Vienna (Maurer et al. 2002, p. 1). In doing this, the interests of the different actors relevant to agriculture and its land were to be factored in (e.g., farmers, urban planning agencies) (Maurer et al. 2002, p. 1). Then, 22 options for Viennese agriculture were developed based on a study of the general conditions relevant to it, and concrete recommendations as to their implementation were introduced. The purpose of these options was to serve as a "basis for decisions for the city administration and as a proposal for Viennese agriculture, in terms of a sustainable development of open landscapes in the city" (Maurer et al. 2002, p. 1, own translation). The project proceeded in two major sections: status quo analysis and option development.

3.5.3.1 Status Quo Analysis

First, an analysis of the current situation of Viennese agriculture was carried out, on the basis of which the options were then developed. Particular topics affecting the then current situation were first presented in a status quo report (Maurer et al. 2002, p. 3): "natural landscape requirements of agriculture", "agricultural production sectors and methods", "economic analysis", "actors and communication", "agricultural land structure" and "city development and land dynamics" (Maurer et al. 2002, p. 3, own translation).

As an example, we document below the content and the fundamental results of the status quo analysis for the area of "natural landscape requirements of agriculture", which was concerned with geomorphology, climate, pedology (soils), and the situation of groundwater and of contamination (Maurer et al. 2002, p. 3).

To ascertain the status quo, the available literature was evaluated and official statistics were used (Kromp et al. 2004, p. 22). Nitrogen balances were calculated for Viennese tillage, horticulture and viticulture (Kromp et al. 2004, p. 22), and project-specific data was collected, for example, on composting and its use (Kromp et al. 2004, p. 22). Existing measurement data were also used, from Municipal Department 45, Hydraulic Engineering, and Municipal Department 22, Environmental Protection, on fertiliser residue contamination (nitrates, pesticides) in groundwater and on air and soil contamination (Kromp et al. 2004, p. 22).

The status quo analysis also served the purpose of changing the perception of Viennese agriculture, for example, in the Viennese municipality departments. In this way, the participating scientists emphasised that Vienna was the capital of Austrian commercial horticulture (Kromp et al. 2004, p. 22) and that agricultural lands, with 13 % tillage, 2 % vineyard and 2 % horticulture areas, were an important share of Vienna's green zones (Kromp et al. 2004, p. 22). Products like sugar beets, cucumbers, tomatoes (Fig. 3.18) and butterhead lettuce dominated agricultural production from the perspective of economic significance, and wheat cultivation from that of area (Kromp et al. 2004, p. 22). Additionally, the results of the status quo analysis reveal facts of which not all Vienna dwellers are aware: "Vienna produces […] incidentally, 90 % of all greenhouse cucumbers as well as […] around 50 % of all tomatoes grown in Austria" (Kromp et al. 2004, p. 22, own translation). During the investigation it also became clear that, in spite of the location being a major city, hardly any environmental pollution was found for agriculture that would be reflected in an increased pollution load in the agricultural

Fig. 3.18 Tomato production in Vienna. *Source* LK Wien, Hammerschmid (*Photo*)

products (Kromp et al. 2004, p. 23). Vienna is deemed to be a "vegetable capital". One third of the vegetables consumed in the city today are also produced in the city itself.[4]

However, agriculture causes environmental pollution through fertilisation (nitrogen surpluses in groundwater) and the use of pesticides (Kromp et al. 2004, p. 23).

Detailed descriptions of the status quo were compiled for the other areas as well, and in 2001 these were documented in the project's first interim report. This had to do, for example, with the situation and economic potential of agricultural direct marketing (Gutheil-Knopp-Kirchwald 2004), with the existing urban planning instruments and their significance for agriculture, and with the function of agriculture for particular aspects of urban planning (e.g., the creation of local recreation areas), which concerns the safeguarding of urban free spaces (Meyer-Cech and Seher 2004, p. 4). In this way, for the first time, there was a stocktaking of the current situation of Viennese agriculture, the results of which then served as a starting point for the development of its options for the future.

[4]www.wien.gv.at/stadtentwicklung/projekte/landschaft-freiraum/landschaft/landwirtschaft/gemuese.html accesed on 19 Feb 2016.

3.5.3.2 Development of Options

The analysis of the status quo was the starting point for the development of options for Viennese agriculture. In this regard, the project's final report says:

> Options are suggestions for measures that should generate more favourable conditions for agriculture at a local level, on the one hand, and should represent possibilities for action for farming businesses themselves, on the other. The goal of the options is an economically viable, sustainable development of agriculture in Vienna. (Maurer et al. 2002, p. 3, own translation)

Based on the situation of agriculture as examined at the time of the start of the project, potential weaknesses and problems were then identified as "trouble areas" (Maurer et al. 2002, p. 4), which affect the future of agriculture in the city (Maurer et al. 2002, p. 4). This was followed by a definition of the possible functions of agriculture in the urban area (Maurer et al. 2002, p. 4). Here, functions can be in the interests of the city (local recreation, landscape variety, groundwater provision) (Maurer et al. 2002, p. 4) or they can be fulfilled by agriculture for the consumers (offer of agricultural products) (Maurer et al. 2002, p. 4).

In addition, an intensive analysis of the conditions for Viennese agriculture was incorporated into the development of the options. Here, influences were examined at an international level (World Trade Organization WTO, eastern expansion of the EU, the joint European agricultural policy), at a national level (Austrian programme for the rural areas) and even at the level of the Viennese framework (e.g., spatial development), and of the key objectives of the relevant municipal departments: Municipal Department 22, Environmental Protection, and Municipal Department 18, City Development and Planning (Maurer et al. 2002, pp. 12 ff.). The goal was to develop the options on the basis of the greatest possible knowledge of general framework and initial conditions and potential future developments.

The options were discussed in a design phase in the context of a workshop with interns. The result of the workshop was the formation of interdisciplinary teams whose task was to develop the options more in detail (Maurer et al. 2002, p. 4). The results of the project were 20 options with different emphases that were distinguished according to who had to take the initiative for their implementation: the City of Vienna or the farming businesses themselves (Maurer et al. 2002, p. 6).

A presentation of all 20 options in this context would go too far. In order to convey an impression of these options as a central project result, below is a concise presentation of one chosen option for each emphasis.

3.5.4 Emphasis: Framework Conditions

3.5.4.1 Example: Development Plan for Agricultural Structures

Among other things, the suggestion here was to work out development plans for agricultural structures for the City of Vienna, given that there was a dearth of coordinated strategies and models for Viennese agriculture. To this effect, the project proposed the use of the instrument called "Development Plan for Agricultural Structures" ("*Agrarstruktureller Entwicklungsplan*", *AgSTEP*), which was known in Germany, but which had not been yet used in Austria (Maurer et al. 2002, p. 64). The content of an AgSTEP is an integrated model for agriculture, in the project for Viennese agriculture, with the individual production sectors (Maurer et al. 2002, p. 7). Strategies and models for agricultural businesses are developed with the aid of development plans for agricultural structures, and are based upon the analysis of initial conditions (Maurer et al. 2002, p. 64). About this, the project's final report says that: "From an agricultural perspective, one must assume a position in respect of the following topics: settlement development, the interests of nature conservation and landscape management, measures in the areas of leisure and recreation, the employment situation outside farming, and production alternatives" (Maurer et al. 2002, p. 64, own translation); and furthermore: "In summary, the following content can be established for an AgSTEP that is desirable for Viennese agriculture:

- an integrated model for Viennese agriculture (for the individual production sectors)
- statements from the agricultural perspective on topics like residential development, recreation, etc." (Maurer et al. 2002, p. 64, own translation).

3.5.5 Emphasis: Production Alternatives

3.5.5.1 Example: Organic Farming

An option in this area affects the production of organically-farmed agricultural products: Although Vienna, as a metropolis, shows a great potential for the marketing of "organic products", the production of these in Vienna is lower than the Austrian average (Maurer et al. 2002, p. 8). Given the geographical closeness to the clientele, Viennese agriculture could more strongly market organic food products here (Maurer et al. 2002, p. 8). A driver for a stronger conversion of production here would be the public domain, since the City of Vienna opts increasingly for the purchase of organically-farmed products in its public institutions (schools, retirement homes, hospitals) (Maurer et al. 2002, p. 8). In addition, Maurer et al. (2002) point out that, in the preschool childcare facilities there are advertisements "[...] for ready-to-serve meals with a minimum proportion of organic products of 30 %, which should be increased to 50 % within 2 years. This results in the possibility for producers in Vienna, as regional purveyors, to convert to organic farming in the future" (Maurer et al. 2002, p. 8, own translation).

3.5.6 *Emphasis: Marketing Alternatives*

3.5.6.1 Example: Regional Marketing

This option contains the recommendation, for farming businesses, that they should bank more on direct marketing. This is because, in spite of the geographical closeness between producers and consumers, this share is still too small. Some of the concrete suggestions for this are farm sales, (organic) farmers' markets, and direct marketing from farmers to consumers (Maurer et al. 2002, p. 9). In doing this, individual businesses could concentrate on fewer products and could address consumers with purchasing power who are willing to pay more for high-quality products, given the situation in Vienna (Maurer et al. 2002, p. 9). In this regard, the project's final report says: "In this way one can not only increase the added value of the individual business, but also the consumers' ties to regional products and the contact between farmers and the urban population can be improved. It is beyond question that these options cannot be realised without the corresponding restructuring within the businesses and collective marketing. A possible first step towards this would be the establishment of an umbrella brand for the products of Viennese agriculture" (Maurer et al. 2002, p. 9, own translation). An umbrella brand implies the development of collective quality criteria for agricultural products, which then farmers have to fulfil if they wish to sell their products under this brand. In this way, agricultural products of high quality could be marketed jointly and professionally.

3.5.7 *Emphasis: Landscape and Local Recreation*

3.5.7.1 Example: Contractual Nature Conservation

With contractual nature conservation, agricultural land can be removed from farming, and nature conservation areas can be created through countryside conservation measures, for example, for the purpose of increasing species variety (Maurer et al. 2002, p. 10). Farmers receive financial compensation and, through the countryside preservation measures that must be carried out, new income sources (Maurer et al. 2002, p. 9). The background was that the then new Viennese Nature Conservation Law provided for this contractual nature conservation for the first time, and that Municipal Department 22, Environmental Protection, was responsible for its implementation.

3.5.8 *Emphasis: Education, Art and Entertainment*

3.5.8.1 Example: Education for Nature and Agriculture

In this option, an increasing "alienation" (Maurer et al. 2002, p. 10) of the urban population from nature and agriculture was to be counteracted through educational projects. The final report mentions ideas like the establishment of a "children's farm", or activities with schools, as a starting point so children can learn something about the origin of food products (Maurer et al. 2002, p. 10). "Doing this should raise understanding for (urban) agriculture, and additionally, by means of these arrangements knowledge about ecological connections can be imparted based on the example of agriculture" (Maurer et al. 2002, p. 10, own translation).

These examples that we have documented from the final report show the intricacy and various starting points of the options. For us the question now became whether in fact practical effects emerged from the options developed in the projects. These and the other options were all developed based on the concrete situation of Viennese agriculture and its surrounding conditions. They were therefore conceived expressly for feasibility.

3.5.9 *The Principal Effect: Enhancement of Viennese Agriculture Through Integration into Urban Development*

What effects resulted from the project "Options for Viennese Agriculture"?
The greatest practical effect, in the eyes of those with whom we spoke about the project, among others, is that in fact, an agricultural structural development plan for the City of Vienna was worked out to immediately stimulate the options (Municipal Department 58, 2004). To this effect the then Councilwoman for the Environment, Isabella Kossina, had initially assigned to a work group the design of an agricultural structural development plan. This work group, in which many people participated, worked between 2002 and 2004 and then submitted a final report, the AgSTEP (Municipal Department 58, 2004). In the agricultural structural development plan those areas where defined which "should be reserved for agriculture in the long term" (Kromp et al. 2004, p. 24, own translation). In addition, the plan contains "propositions for measures which should contribute to a long-term safeguarding and preservation of the cultivation of these demarcated areas" (Kromp et al. 2004, p. 24, own translation). The share of the priority areas in the AgSTEP relative to the Viennese agricultural area as a whole is 69 %; that corresponds to more than 4800 ha (Kromp et al. 2004, p. 24). Proposals for the development of the agricultural businesses in Vienna were prepared in order to secure the agricultural management of these lands in the long term (Kromp et al. 2004, p. 25).

3.5.10 Goals of AgSTEP

The AgSTEP for Vienna should

- include a demarcation of those areas that should be reserved for agriculture in the long run, as a complement to the goal definitions for the green area planning (so-called "Greenbelt 95").
- be considered in the next city development plan (STEP 05).
- develop measures and recommendations, as a complement to this, that would contribute to the long-term safeguarding of the cultivation of these demarcated areas (Municipal Department 58, 2004, p. 5).

One recommendation of the work group of the agricultural structural development plan was to incorporate this into the Viennese urban development plan "STEP 05" (Stadtentwicklung Wien 2005), which was to be set up in 2005. In fact, spatially relevant aspects of the agricultural structural development plan were then incorporated in the STEP 05 (Meyer-Cech and Seher 2004, p. 4) (see Fig. 3.19). In particular, the designation of four land categories in respect of the future agricultural use is remarkable here, because this originated directly in the ideas of the AgSTEP (Meyer-Cech and Seher 2004, p. 4). In the STEP 05, priority areas for agriculture really are designated according to the recommendations of the AgSTEP. Concerning this, the urban development plan says:

> The total extension of the land designated as 'agricultural priority area' is 4827 ha, which is 69 % of the total farmland in Vienna. This is not a precise demarcation by plots, but rather a rough demarcation of areas based on natural, urbanistic and infrastructural circumstances. Areas in categories 1 and 3 should remain reserved for agricultural use in the long run. (Stadtentwicklung Wien 2005, p. 143, own translation)

The documentation for the results of the Cultivated Landscape Research Programme (*Forschungsprogramm Kulturlandschaftsforschung*, KLF) says the following about this:

> The municipal administration and the agricultural sector of the borough of Vienna together lay down the 'agricultural structural development plan' and incorporate it into the Urban Development Plan 2005. The key objectives of the 'agricultural structural development plan' are the preservation of the cultivation of the agricultural areas of Vienna in accordance with the green space development plans, and to make it more ecological. In this way, for the first time agriculture in Vienna will go beyond mere declarations of intent and will be incorporated into the city planning (bm:bwk 2005, unpaged, own translation)

The urban development plan is not binding and only has the status of a directive that must be concretised through more planning (Meyer-Cech and Seher 2004, p. 4). Nevertheless, those participating in the project at the time emphasise the great practical significance of this process: According to Maurer this is that before, the agricultural lands had no particular political significance, they were simply not a topic for the city planning (Prof. Maurer, then at the Ludwig Boltzmannn Institute for Organic Farming, personal communication, July 22, 2010). From the point of view of the Viennese Chamber of Agriculture, the needs of the agricultural sector

Agricultural Structural Development Plan for Vienna (AgSTEP)

Safeguarding agriculture in Vienna

Agriculture was treated as a distinct sector, one separate from the land utilization demands of other economic sectors, even before the creation of STEP 05 as part of the agricultural structural development plan. The results for the long-term safeguarding of agriculture in Vienna achieved this way will be adopted into STEP 05 and will be coordinated with other development goals.

Development goals

The long-term goals for Viennese agriculture are **the preservation (safeguarding) of the cultivation of the areas used for farming and the further expansion of an environmentally sound production** – in accordance with the long-term goals of green area planning (…).

To this end, the 'Agricultural Structural Development Plan for Vienna' (AgSTEP, for *Agrarstrukturelle Entwicklungsplan*) was developed together with the Viennese Chamber of Agriculture. It contains:

- demarcation of the areas which shall be reserved for agriculture and
- proposals for measures that should contribute to a long-term safeguarding and preservation of the cultivation of these demarcated areas

The total extension of the land designated as 'agricultural priority area' is 4,827 ha, which is 69% of the current total farmland in Vienna. This is not a precise demarcation by plots, but rather a rough demarcation of areas based on natural, urbanistic and infrastructural circumstances. Areas in categories 1 and 3 should remain reserved for agricultural use in the long run.

Some of the proposals emphasising the long-term cultivation of the designated agricultural priority areas and suggestions for the development of farming businesses in Vienna are reproduced here as excerpts.

Fig. 3.19 Excerpt from the STEP 05 (integrates AgSTEP). *Source* Stadtentwicklung Wien (2005, pp. 142–143), own translation

were able to be articulated and introduced (information from Mr. Windisch, President of the Viennese Chamber of Agriculture, personal communication, July 13, 2010). The Chamber of Agriculture itself was not necessarily enthusiastic at the beginning of the options project, it must be said. This had to do with the fact that there were different opinions among the farmers as to what was to happen with the lands—some were interested precisely in selling, with the prospect of profit from the repurposing into building plots. However, since the farmers had the possibility of participating in the project from the beginning, and of articulating their needs, a rapprochement could definitely be achieved around difficult topics like the expansion of organic farming or nature conservation (information from Mr. Windisch, President of the Viennese Chamber of Agriculture, personal communication, July 13, 2010).

In conversation, Mr. Kubik (Municipal Department 22, Nature Conservation) stressed that the results that influenced the STEP 05 in relation to agriculture are today applied in the day-to-day work of the Department, as before (personal communication, July 20, 2010).

With the help of the "options" project, agriculture attained a strong valorisation; it now belongs increasingly in the public perception to Viennese identity. The idea of a 3-zone concept became a part of the urban development plan. The AgSTEP is now, and at least until 2015, a part of the urban development plan (Stadtentwicklung Wien 2005): The current STEP 05 urban development plan is valid for that period.

Along with this spectacular impact, we could also establish other effects of the project:

One option was the expansion of the production of organically-farmed agricultural products. Kromp, Hartl and Maurer showed, in an article in 2004, that the share of "organic and transition businesses [...] more than doubled by 2003, and the area more than trebled, at 840 ha or 14.3 %" (Kromp et al. 2004, p. 23, own translation).

In addition, concrete measures were implemented in practice, to which the options also referred: The Vienna city council committed to buy organic products for at least 30 % of its food purchases for its public facilities, and it has set up stands for organic farmers at the Viennese markets (Kromp et al. 2004, p. 23).

The contractual nature conservation programme called *Lebensraum Acker* ("field habitat") was implemented in cooperation with Municipal Department 22, Nature Conservation. There was a pilot phase for this starting in 2000, before the programme was started in 2002 (Kromp et al. 2004, p. 24). At the beginning of the project existing conflicts between the agricultural sector (Viennese Chamber of Agriculture) and nature conservation interests were able to be diminished not least because of the contractual nature conservation option, for which there was close collaboration with the Viennese Chamber of Agriculture. Mr. Windisch, who was then Vice President and is now President of the Viennese Chamber of Agriculture, indicated in an interview that the options report represented the start of the interaction between nature conservation and Viennese agriculture (personal communication, July 13, 2010). He describes this process as one in which nature conservation representatives as well as farmers learnt from one another, and in which the situation went from being one where nature conservation was seen as an enemy in the eyes of farmers to one of increased mutual acceptance, through de-escalation (personal communication, July 13, 2010).

Yet another effect is that the cooperation between the Chamber of Agriculture and the City Council resulted in different marketing initiatives: Since 2001 there is a harvest festival in Vienna in which Viennese agriculture can expose its product diversity. Various quality and image campaigns were begun for Viennese products, for example, Viennese wine. In this case, the event of the *Wiener Weinfrühling* (Viennese wine spring), which takes place every year in March and April with festivities and degustation, established an initiative for the marketing of wine (Mr. Windisch, Viennese Chamber of Agriculture, personal communication, July 13, 2010). The Viennese Chamber of Agriculture started a campaign called "we plough for Vienna", so that consumers come to prefer Viennese agricultural products, and in order for the significance of Viennese agriculture to reach the consciousness of the people. However, the establishment of a Viennese umbrella brand (included as an option) could not be realised.

The exhibition, "Agriculture in Vienna", the content of which the researchers prepared on behalf of the Municipal Department 18, Urban Planning, was also conceived in the context of the project (Maurer et al. 2002, p. 126). The goal of the exhibition was to show what agriculture in a major city like Vienna looks like. The exhibition was shown in three places: in the Viennese planning workshop in 2001, in a mall in 2002, and also in 2002 as part of the *Strohzeit Favoriten* (*Strohzeit*, in English "straw time", is an Austrian festivity, and *Favoriten* are "the favourites") (Stadtentwicklung Wien 2002, p. 5). The final report about this says the following:

> From September 27 to November 9, 2001, the extent and significance, but also the problems, of urban agriculture were presented in the Viennese Planning Workshop. This described a number of topics that ranged from the variety of crops grown in Vienna and the consumption of particular foods by the city population, to the different functions of agriculture in the urban area, and also included the economic difficulties of agriculture. The target audience was the interested public, particularly schoolchildren. (Maurer et al. 2002, p. 126, own translation)

Guided tours were offered in the context of the exhibition, as well as excursions to farms. A total of 700 schoolchildren participated in the guided tours for school classes (Maurer et al. 2002, p. 126). Brigitte Jedelsky, who was then a city planner in the Municipal Department 18, Urban Development, explained to us in conversation that one of the important functions of this exhibition was to "approach" the population with topics in city planning, and that this is normally very difficult with such cumbersome topics (personal communication, July 12, 2010).

3.5.11 *Factors for the Success of Knowledge Transfer*

3.5.11.1 Transfer Through External and Internal Allies

Various allies were important for the success of the knowledge transfer in the options project. These played different roles as external and internal allies. First of all, one must mention the different Viennese municipal departments participating in the project. In contrast to other allies presented in this book, we call these "internal allies", in this case because, firstly, they acted as co-financiers. The different municipal departments participating in the project were themselves integrated in the research process, particularly in the development of options. They had an interest in obtaining information about the role of agriculture in their respective fields of work (for example, Municipal Department 18, Urban Development and Municipal Department 22, Nature Conservation), and they supported the project with their own collaboration, especially in the development of options in the work group intended for that purpose, and by making their own data available. In this way, their collaboration was no longer only external; they were participating internally in the project, and therefore as allies, a guarantee that the project results would influence their own work later, be it in the urban development planning (Stadtenwicklung Wien 2005) or in the measures for contractual nature conservation, which the Municipal Department 22 carries out.

Another example of an internal ally is the Municipal Department 49, Forestry and Farming. It had an interest in this project because it wanted to promote agriculture within city policy: It had to be shown, on a scientific basis and beyond discussion, that agriculture has great significance for Viennese urban development. First, the comprehensive presentation of data on Viennese agriculture was important for this purpose. The interest in clearly showing the significance of agriculture to other actors and to the citizens was something the Municipal Department 49 shared with the Viennese Agriculture Chamber. As a result of this clear interest in the research project, the researchers could also work on topics that, at first, were not in accordance with the notions of the agricultural actors. Questions concerning nature conservation and the expansion of organic farming became a big part of the project as options: Both topics were seen critically at first by the Agriculture Chamber; however, the prospect of the desired project content with its results overlaid the negative attitude towards the more undesired topics.

The project achieved results that were of direct interest for agriculture. Since the interest in a strengthening of the position of Viennese agriculture was so great, the rejection of the undesired project content was significantly less. Furthermore, without the cooperation with the Agriculture Chamber, the project could not have had any success; this is because many of the options of the project could be implemented only via the farmers' business decisions. The Agriculture Chamber, which was not particularly receptive at the beginning of the project, changed its position here and, as an ally, it was able to influence the farming businesses, so that the ideas resulting from the options could be implemented.

With regard to nature conservation, there were changes even in the position of the Agriculture Chamber as an internal ally. One example is contractual nature conservation. Even the initial, strongly negative, position regarding an expansion of organic farming became weaker in the course of the project. If the Agriculture Chamber has a positive attitude towards the project and its options, then the behaviour of the individual farming businesses also changes. The Agriculture Chamber was involved in the project directly, for example, in the development of the options. It became an internal ally.

But there was also an *external* ally: The Councilwoman for the Environment, Isabella Kossina, appointed a work group in 2002 to develop the agricultural structural development plan for the City of Vienna. With this, there was a "government" assignment to actually put the project results into practice. A political actress, the Councilwoman for the Environment, in her role as external ally, committed other actors to implement the options in accordance with the formulation of an AgSTEP.

3.5.11.2 Research

In our RIU model of successful knowledge transfer, the research underlying said transfer in practice must be of good quality and should cover the most current stage of scholarship on this knowledge. In the case of the options project, all necessary data on the actual conditions of Viennese agriculture were first compiled. Some of the data were generated scientifically within the project. In the project there was, on the one hand, the exhaustive status-quo analysis of scientific quality, and on the other a scientific description of the conditions of Viennese agriculture, in order to develop the options on the basis of the greatest possible available knowledge. Scientific methods and approaches were developed that corresponded to the most current knowledge. Various scientific partners from different disciplines cooperated in the project in order to be able to process the different research dimensions. The Ludwig Boltzmannn Institute for Organic Farming and Applied Ecology, Vienna, the Institute for Spatial Planning and Rural Development of the University of Natural Resources and Life Sciences, Vienna, the Institute for Interdisciplinary Research and Continuing Education, Department of Social Ecology (*IFF*) of the University of Klagenfurt, and the Institute for Finance Studies and Infrastructure Policy of the Technical University of Vienna worked together.

An international workshop was undertaken, "Agriculture in Vienna—Between Urban Planning and Economics", in which the project participants took part (municipal departments of the City of Vienna, the Viennese Agriculture Chamber and researchers) (Maurer et al. 2002, p. 126). It included talks by foreign experts (Germany). "The goal of the event was to discuss the role of urban agriculture, involving foreign experts—with administrative departments of the City of Vienna, representatives from agriculture and scientists, with regard to the research goal of the project (the development of options for agriculture in Vienna)" (Maurer et al. 2002, p. 126, own translation). The results of the workshop were published separately through the Municipal Department 18, Urban Development, Vienna, which organised the workshop (Stadtentwicklung Wien 2002).

In the development of options, solutions were sought that had already been used outside Austria: The instrument, "agricultural structural development plan" was "imported" from Germany, as several interviewees underscored.

3.5.11.3 Integration

Integration represented an important success factor for knowledge transfer in the options project. The project was orientated towards public welfare goals: One of its major goals was change towards more sustainable agriculture. This corresponded to the societal discussions concerning a modified agriculture, which were current then and now (e.g., organic farming of food products). In addition, the project aimed towards being relevant for different target groups and current decisions in city politics, from the start. This is because the Viennese Municipal Department, as co-financier, expected relevant results in respect of agriculture for its various urban development political goals. The project developed concrete options orientated towards a later implementation, which then could be used by the corresponding municipal departments, as shown: The results were helpful in the coordination between agriculture and nature conservation, and in the further urban development planning, to mention only the most pertinent examples. The practical basis, geared towards relevance from the start, can be recognised as an important success factor for knowledge transfer in this project. There was also relevance with regard to the Viennese Agriculture Chamber. It had the opportunity to articulate its requirements and wishes in the course of the research process, and was integrated commensurately as an important ally for a later implementation of the results. The project was important for the Agriculture Chamber because, for the first time, the significance of agriculture in Vienna was analysed comprehensively on a scientific basis, and there were proposals for conflictive areas, like nature conservation, which were useful for both nature conservation and agriculture (contractual nature conservation).

Precisely this sophisticated analysis of the agricultural situation regarding the different questions in agriculture saw to it that the requirements of practice for science-based solutions were in fact satisfied by means of scientific answers.

The inclusion of the different practice actors in the transdisciplinary research process from the beginning was an important building block within integration. “Along with the interdisciplinarity of the project, special attention was paid to transdisciplinarity, a connection to practice, and to public relations” (Maurer et al. 2002, p. 1, own translation). In addition, there was an advisory board that was organised by the Municipal Department 22, Environmental Protection. The members of this board were: the Agriculture Chamber, Municipal Departments 18, 21c, 22, 40, 49 and 58 (City Development and Planning, District Planning and Land Use, Environmental Protection, Technical Land Matters, Forestry Office and Farming Businesses, and Legal Matters of Cultivation, respectively), the Viennese Nature Conservation Advocacy, and planning offices (Maurer et al. 2002, p. 1). The board was informed regularly on the progress of the project and had the assignment of delivering statements (Maurer et al. 2002, p. 2). Through the work of the board, different sections of the city’s administration could be increasingly sensitised regarding agriculture and its significance and concerns (Maurer et al. 2002, p. 2). The following examples substantiate the significance of the advisory board for integration: Agricultural aspects from the perspective of the Viennese city administration were introduced in the first session, which took place whilst still in the proposal phase of the project, in 1999 (Maurer et al. 2002, p. 125). "Viticulture", as a topic, became a subject of the project only after encouragement from the practice side—through this example one can very well see the significance of integration: Scientific questions are developed from the concrete needs of practice. The end result, the options, were discussed critically in the fifth session of the project's advisory board; here, the Viennese Agriculture Chamber expressed reservations concerning certain proposals. In addition, the Agriculture Chamber remarked that there was a lack of options in certain areas (Maurer et al. 2002, p. 125).

Various media and forms of presentation were also used in the options project according to the specific target groups. Examples of this are the "Agriculture in Vienna" exhibition, mentioned above, which had "the greater public" as a target group, and the guided tours and excursions to farming businesses that were on offer, for which the target group was "schoolchildren" (Maurer et al. 2002, p. 2). For these, the project researchers carried out the integration work for the content conception.

The international workshop in 2001 brought researchers, the city administration, and international experts together: In this context, a panel discussion in the ballroom of the Vienna City Hall was also organised (Maurer et al. 2002, p. 126).

In all, in the options project there was an integration of practice from the beginning, which according to our dialogue partners was also strengthened due to the importance given to a language suitable in general and for practice, and to the processing of the project results.

3.5.11.4 Utilization

The options project resulted in concrete utilization products for different target groups: There was a well-attended exhibition that was presented several times (the target group was the townspeople, in particular schoolchildren), numerous publiccations, for example, one on the topic of "urban agriculture" in the journal *Land & Raum* (Land and Space) (3/2004), which contained articles on the project results (target group: expert audiences). A publication, "Agriculture in Vienna—Between Urban Planning and Economics", resulted from the international workshop, as "Workshop Report No. 52" of the Vienna City Development Office (Municipal Department 18) (target group: expert audiences).

The most important utilization products, however, were the concrete options for Viennese Agriculture, which were presented in a final report suitable for practitioners (Maurer et al. 2002), which was delivered to the practice partners (city administration, contracting entities, Agriculture Chamber) (target group: practitioners). As our analysis showed, being the most important utilization products, these were the initial force for the development of an agricultural structural development plan, and for the inclusion of the latter in the STEP 05 urban development plan (Stadtentwicklung Wien 2005). The project's utilization products became part of the greater Viennese urban development planning.

3.5.12 *Knowledge Transfer Success Factor List*

The following figure (Fig. 3.20) shows that all the main criteria for RIU were fulfilled in this process. If a particular criterion is of special relevance, it is marked "+".

Options for Viennese Agriculture
Areas of activity for knowledge transfer (RIU Model)

1	Research	
■	Assessing current scientific information	✓ +
■	Compliance to the procedures of good scientific practice	✓
■	Cooperation with other scientific institutions and projects	✓
■	Independent meaningfulness of scientific findings	✓

2	Integration	
■	Orientation toward public goals (like sustainability, …)	✓
■	Relevance in regard to political processes	✓ +
■	Relevance in regard to allies (external and internal allies)	✓ +
■	Target-group oriented intermediation	✓ +

3	Utilization	
■	Contribution to democracy	✓
■	Contribution to the rule of law (constitutional legality)	✓
■	Contribution to "good governance"	✓
■	Appropriate solutions to problems	✓
■	Participation in the scientific discourse	✓

Fig. 3.20 Success factor list for the options for Viennese agriculture

3.6 Establishing a Legal Basis: Implementing the EU Water Framework Directive in Austria

3.6.1 Project Fact Sheet

- **Project**

River landscape types in Austria
Leitbilder (guiding visions) for a sustainable development of river landscapes
Other different projects

- **Type of project**

Research project

- **Participating research institutions**

  - Department of Hydrobiology, Fisheries and Aquaculture of the University of Natural Resources and Life Sciences, Vienna (*Institut für Wasservorsorge, Gewässerökologie und Abfallwirtschaft der Universität für Bodenkultur Wien*)
  - Institute of Ecology and Environmental Planning, Klagenfurt (*Institut für Ökologie und Umweltplanung Klagenfurt*)
  - Institute for Sustainable Economic Development, University of Natural Resources and Life Sciences, Vienna (then *Institut für Wirtschaft, Politik und Recht der Universität für Bodenkultur Wien*)
  - Austrian Institute for Sustainable Development (*Österreichisches Institut für Nachhaltige Entwicklung Wien*)
  - Office for Spatial and Regional Planning Vienna (*Büro für Raum- und Regionalplanung Wien*)

- **Location**

Different locations in Austria

- **Central question**

What types of river landscape exist in Austria, and what are their natural characteristics?

- **Goals**

  - Creation of an Austrian typology of running waters based on their natural ecological potential
  - Description of their ecological quality according to the EU Water Framework Directive

- **Timeline**

2000-2003

- **Financing**

  - Cultivated Landscape Research (*Kulturlandschaftsforschung*, KLF), sponsored by the then Austrian Federal Ministry of Education, Science and Culture (*Österreichisches Bundesministerium für Bildung, Wissenschaft und Kultur*)
  - Austrian Federal Ministry of Agriculture, Forestry, Environment and Water Management (*Österreichisches Bundesministerium für Land- und Forstwirtschaft, Umwelt und Wasserwirtschaft*) (different projects for the implementation of the EU Water Framework Directive in Austria

3.6.2 Background: The EU Water Framework Directive

Austria became a member of the European Union in 1995. Since then, Austria has the obligation, like all member states, to formulate national law that corresponds to the regulations formulated by the European Union. Environmental policy is the policy field with the most strongly evolved European dimension. The European Union has introduced a huge number of regulations that demand from the members the implementation in their national law (Böcher and Töller 2012, p. 163). An example of such challenging national activity in lawmaking is the EU Water Framework Directive. In this chapter we show the important role scientific knowledge has played in this legal task.

On 22 December 2000 the EU Water Framework Directive became legally binding for all member states (Fig. 3.21). The goal of the Directive was to create a coordinated and coherent water policy within the European Union. The quality of lakes and rivers should have reached an acceptable standard by 2015 and a decline in water quality should be hindered (Bundesministerium für Land- und Forstwirtschaft, Umwelt und Wasserwirtschaft 2006, p. 4).

Austria is a small country of 83,855 square kilometres, but it has a great number of lakes and rivers. The combined length of all its greater rivers that have a drainage basin of more than 10 km^2 each is about 31,000 km. The total length of all rivers in Austria is about 100,000 km. Additionally, Austria has 6000 lakes, out of which 30 have an area greater than one square kilometre (Bundesministerium für Land- und Forstwirtschaft, Umwelt und Wasserwirtschaft 2011, p. 9) (Fig. 3.22).

These numbers demonstrate that water streams are of great importance for Austria. Rivers and lakes are ecosystems, water supplies and traffic lanes, but they are also areas for recreation and offer a potential for tourists and the economy. The ecological quality of the lakes and rivers has reached the standard ratings of "good" and in some cases "very good" (Bundesministerium für Land- und Forstwirtschaft, Umwelt und Wasserwirtschaft 2007, p. 11). This high standard was met recently

22.12.2000 EN Official Journal of the European Communities L 327/1

I

(Acts whose publication is obligatory)

DIRECTIVE 2000/60/EC OF THE EUROPEAN PARLIAMENT AND OF THE COUNCIL

of 23 October 2000

establishing a framework for Community action in the field of water policy

Fig. 3.21 Excerpt of the original document: The EU Water Framework Directive, published in the Official Journal of the European Communities. *Source* Official Journal of the European Communities L327/1, 22 Dec 2000

only as the result of active policy and various technical, economic and political interventions for the improvement of water quality during the 1960s and 1970s (Umweltbundesamt 2001, p. 196). Basically, the EU Water Framework Directive supports this policy, with the goal of achieving an ecologically healthy water source for life, ecology and recreation. The quality of rivers, lakes and ground water should be protected permanently, within each country and beyond the borders.

Excerpt from the EU Water Framework Directive

Water is not a commercial product like any other but, rather, a heritage which must be protected, defended and treated as such (Directive 2000/60/EC of the European Parliament and of the Council of 23 October 2000 establishing a framework for Community action in the field of water policy, p. 1).

The advantages of the EU Water Framework Directive are the cross-border concept and a new approach to river basin management. A German Guideline for the implementation of the EU Water Framework Directive formulated the key concept well: "Rivers, lakes, coasts, ground water and their flood areas, as well as their drainage basins are looked at as one comprehensive unit for river basin

Fig. 3.22 Recreation by water sports at the Lake Neusiedl, federal state of Burgenland, Austria. *Source* Michael Böcher (*Photo*)

management. All technical means should optimise the whole area and should not stop at administrative borders or zones of legal jurisdiction anymore. Only the natural conditions define the comprehensive units for water management. These units for each river comprising the ground water and the coasts are the framework used to coordinate all technical and economic means of river basin management. All member states of the European Union have to create administrative structures to implement this comprehensive concept in practice" (DVL 2010, p. 8, own translation).

The goal is to maintain the lakes and rivers as "healthy and functioning ecosystems and to guarantee a sustainable protection of the source water" (DVL 2010, p. 9, own translation). In principle, the quality of water should not decline and all water areas that do not meet the adequate quality level should be improved (DVL 2010, p. 9). Exceptions are allowed only within water areas that have been used for many decades. For example, in these areas shipping traffic is allowed even if it pollutes the ecosystem, if shipping is indispensable for maintaining economic activity (DVL 2010, p. 9). The member states have the option of enabling shipping traffic by using specific labels that are granted by the European Union (DVL 2010, p. 9).

The member states have an obligation to work out River Basin Management Plans (RBMPs) for each river basin district. The relevant organisations and the public have to be informed and involved (Bundesministerium für Land- und Forstwirtschaft, Umwelt und Wasserwirtschaft 2006, p. 4, p. 8).

3.6.3 Implementation of the European Water Framework Directive in Austria

Austria planned the implementation of the European Directive in several steps until 2015. Austria shares the drainage basins of the three biggest European rivers, the Danube, Rhine and Elbe (Bundesministerium für Land- und Forstwirtschaft, Umwelt und Wasserwirtschaft 2006, p. 8). The multiple activities should follow a stepwise plan (Table 3.5).

The plan makes visible the huge number of activities at the national level. It started with the transformation of the European Directive into national law, which was done in Austria in 2003. An evaluation of the status quo of the rivers and lakes had to follow the final results of characterising all rivers and lakes by specific types. The River Basin Management Plans (RBMPs) were to be made based on this information. After public involvement, the final versions of the management plans were to become a legally-binding regulation by the Austrian Federal Ministry of Agriculture, Forestry, Environment and Water Management. This important step

Table 3.5 Step-wise implementation of the EU Water Framework Directive in Austria

Date	Activities in Austria
2003	Formulation of national law based on the EU Water Framework Directive
March 2005	Report on the monitoring of the quality of rivers and lakes in Austria; typology of rivers and lakes and ecological and economic analysis
December 2006	Programmes for the improvement of monitoring and material for public information
December 2008	Draft of River Basin Management Plans (RBMPs) and public involvement
December 2009	Decision on River Basin Management Plans (RBMPs)
By 2010	Political programme for securing the efficient use of water
December 2012	Establishing a new monitoring and evaluation system
December 2015	Interim report on the implementation of the EU Water Framework Directive

Source Bundesministerium für Land- und Forstwirtschaft, Umwelt und Wasserwirtschaft (2006, p. 9) (own translation)

was reached in 2009. The whole legal package became legally-binding by being published in the Codex of Austrian law on 30 March 2010 (NGPV 2009).

The national River Basin Management Plan comprises the rivers Elbe, Danube, Drau, March (including the Thaya), Mur, Rabnitz, Leitha and Rhine (NGPV 2009, §2). For these areas, the European Water Framework Directive's principles and procedures became binding. An important portion of the rivers were classified under the type "heavily modified and artificial bodies of surface water" (NGPV 2009, §3) which means that they fulfil the conditions for exceptions devoted primarily to commercial use. As a consequence, the general obligation for ecological improvement is not required politically in those areas anymore.

Other parts of rivers and lakes are considered fit for a stepwise ecological improvement (NGPV 2009, §4). The national plan shows the recent stage of rivers and lakes and fixes an obligatory schedule as to when the ecological standards should be met (NGPV 2009, §4). For some rivers in bad condition, the plan leaves the method used to improve the ecological quality realistically open (NGPV 2009, §4). This illustrates the huge technical and political challenge of ecologically improving the multiple uses of rivers and lakes that provide products and services for many people who make their living from lake and rivers.

Despite exceptions, the national guidelines formulate new and relevant political instruments and technical means to improve the ecological standards of rivers and lakes and of groundwater. The water management has to protect, use and develop the water areas in such a manner that preserves the ecological quality and the amount of water and so that a decline in quality is prohibited (NGPV 2009, §6 ff.).

The overview already makes clear that the implementation of the European Water Framework Directive was a challenging policy task for Austria that needed many years to produce results. This process presupposed that there would be sufficient and high quality information about the use and quality of lakes and rivers in Austria. This was the challenge for science. Scientific knowledge was transferred successfully into the political process and contributed to the implementation's keeping to its time schedule.

3.6.4 The Scientific Project

Providing knowledge for the implementation of the EU Water Framework Directive in Austria was not the task of a single scientific project. Instead, a bundle of research projects was available, which dealt with specific aspects of the management of rivers and lakes in Austria (Muhar et al. 1996, 1998). The sixth environmental report of the Environment Agency Austria (UBA) published a list of eight projects that were highly relevant to river basin management:

> Since 1980, attempts to define types for waters in Austria have been important (…). The following eight projects provided key results that were relevant for the implementation of the European Water Framework Directive:
>
> - Basic concepts and data for the typology of running waters in Austria (Moog nd Wimmer 1990)
> - Methods for the evaluation of ecological functions (Österreichisches Normungsinstitut 1997)
> - Register of the stream order of the Austrian waters (Wimmer and Moog 1994)
> - Flow model of the rivers in Austria (Mader et al. 1996)
> - Monitoring of habitat integrity of running waters (Muhar et al. 1996, 1998)
> - Natural areas of running waters in Austria (Fink et al. 2000)
> - Monitoring of the ecomorphology of running waters
> - Management concept for rivers and lakes. (UBA 2001, p. 232), own translation.

Within the implementation, the Austrian Federal Ministry of Agriculture, Forestry, Environment and Water Management made use of selected knowledge provided by different projects. In addition, in specific important and unsolved questions the ministry initiated and financed new and innovative scientific projects.

One prominent example we are looking at is the project "River landscape-types in Austria—'Leitbilder' (guiding visions) for a sustainable development of river landscapes", which was sponsored by the Austrian Cultivated Landscape Research (KLF) program. Prof. Muhar, a specialist for river ecosystems of the University of Natural Resources and Life Sciences, coordinated the project. The project was built on four modules (1) Typology of ecosystems of rivers, (2) Models for sustainable river ecosystems, (3) Case study of a sustainable river, the Möll in Carinthia and (4) Strategy for implementing models of sustainable rivers (www.flusslandschaften.at).

The first module based the typology on the ecological data and on the data on the multiple uses of the rivers (www.flusslandschaften.at/?s=5&t=1#ob). It was the first time that such data were collected in the field and delivered by science. They were the basis for important tasks in the implementation of the EU Water Framework Directive (www.flusslandschaften.at/?s=5&t=1#ob):

> Such data, especially in light of the Water Framework Directive, which is compulsory for all EU member states and therefore for Austria, are a key prerequisite in order to:
>
> - determine and define characteristic reference stretches
> - define type-specific Leitbilder based on principles of aquatic ecology
> - comparatively evaluate how anthropogenically induced changes reduce the ecological integrity of running waters
> - develop type-specific measures tailored to the respective water bodies (e.g. water management measures, regeneration as well as protective measures)—that focus both on the natural potential and the current status quo—and
> - evaluate their effects
>
> Source: http://www.flusslandschaften.at/en/?s=5&t=1#ob.

A more detailed description of the project work is given by the project itself (www.flusslandschaften.at):

> Considering these highly relevant research policy issues, and based on those Austrian rivers with a catchment area >500 km^2, the project members elaborated the following topics within this sub-module:
>
> 1. Characteristics of the natural/ecological potential of Austria's river landscapes (in the sense of their type-specific, anthropogenically undisturbed condition) based on physical and biological parameters (e.g. morphological river type, discharge conditions etc. as well as fish regions and floodplain vegetation)
> 2. Representation of the ecological status quo based on two selected biological indicators (fish fauna, floodplain vegetation)
> 3. Representation of the uses of and impacts on the rivers and their surroundings based on selected types of impact (e.g. water diversion, damming, uses at the floodplain level etc.)
> 4. Identify river types according to natural landscape and ecological criteria
> 5. Identify available river landscape types as a synopsis of the status quo of the running waters and of the natural potential of the aquatic landscapes.
>
> These results were used to develop Leitbilder for river landscapes in sub-module 2 and then specifically applied to the case study Möll, Carinthia in sub-module 3.
>
> Source: http://www.flusslandschaften.at/en/?s=5&t=1#ob.

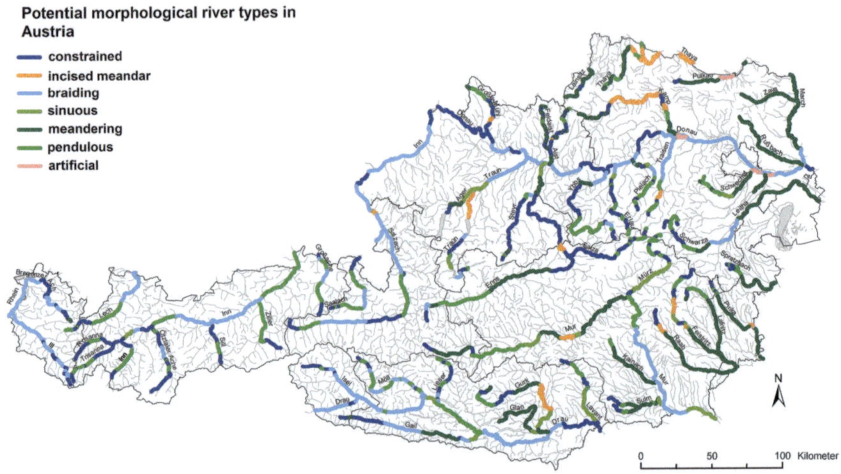

Fig. 3.23 Map of ecological river landscape types in Austria. *Source* Muhar et al. (2003, p. 116)

One of the important results was the classification of rivers based on their ecological type (see Fig. 3.23). These data should provide a scientific and objective basis for an evaluation of rivers following the European Directive. The evaluation identified the specific parts of the rivers that are still close to their natural state and are of high ecological value (information from Ms. Ofenboeck, Austrian Federal Ministry of Agriculture, Forestry, Environment and Water Management, personal communication, July 15, 2011). The required analysis of risks and water stress build on this data. The basic orientation for the typology is how close the current state of rivers compared to their ecological potential. The empirical data enabled the scientists to classify all rivers in Austria according to their type in conformity with the European Water Framework Directive.

3.6.5 The Effects

The results of the project caused a major impact in providing a professional basis for the regulation of river basin management by the Austrian Federal Ministry of Agriculture, Forestry, Environment and Water Management. Such direct effects of science on politics are rare but not impossible (Pregernig and Böcher 2012, p. 209). They fit the model of "linear scientific knowledge transfer", which means that scientists produce scientific knowledge and the politicians understand and use this information as a basis for action (see Chap. 4). During the entire project the

Austrian Federal Ministry of Agriculture, Forestry, Environment and Water Management and the scientists were in close contact. The ministry financed the project and was very interested in the results. As soon as the science-based typology of rivers was achieved, the ministry applied it as a basis of the regulation.

In this extraordinary case an additional structural factor is important. The strong impact was not produced by one huge integrated project but by many different smaller projects that focused on specific questions and were conducted in part a decade before the knowledge transfer began. The individual projects focused on detailed questions on water management and were only loosely connected to each other. The Ministry preferred to pick relevant results from different projects and put them together in a synthesis that was orientated to satisfy the requirements of the European Directive. This approach worked well and the Ministry had no need for an overall scientific synthesis. It made use of the existing project results and added some new aspects by initiating new small projects that produced the needed knowledge more quickly than a huge integrated project could have done. The public administrator of the Ministry, Ms. Ofenboeck, emphasised that she obtained the scientific advice she was looking for and that she could use the typology directly to formulate the technical part of the guideline (Ms. Ofenboeck, Austrian Federal Ministry of Agriculture, Forestry, Environment and Water Management, personal communication, July 15, 2011). The principal scientist, Prof. Muhar of the University of Natural Resources and Life Sciences Vienna, confirmed that the scientific results were used properly by the administration (Prof. Muhar, University of Natural Resources and Life Sciences Vienna, personal communication, October 19, 2011; Muhar et al. 1996, 1998). Of course, the administration selected the results and added its political evaluation in order to formulate the national River Basin Management Plan (RBMP) (NGPV 2009).

This example of adherence to the law shows how the scientific basis was transformed into a legally binding regulation according to the European Water Framework Directive. The scientific basis is the ecological potential and the monitoring of the recent state of the ecosystem. Table 3.6 shows an example from the list that comprises all rivers of Austria and became part of the legally binding regulation (Table 3.6).

3.6.6 Factors for the Success of Knowledge Transfer

3.6.6.1 Transfer Through External and Internal Allies

An important external ally for the research in the implementation of the EU Water Framework Directive is the European Union. By formulating the binding European Water Framework Directive, the European Union created an obligation for all member states to carry out comprehensive monitoring of their rivers and lakes and to evaluate them based on scientific criteria and data. The pressure of the European Union caused a demand for scientific information on water management within the

Table 3.6 Strongly influenced rivers according to use (sample of the attachment of the NGPV2009, only available in German language)

Tab. FG-erheblich-veränderte-WK: Fließgewässer - erheblich veränderte Wasserkörper

Detail-OWK-Nummer	betroffene Bundesländer	Fluss	Fluss-km (von)	Fluss-km (bis)	Zustand Biologie bezüglich natürlicher Referenzzustand	Charakter verändert durch:					Sanierungsmaßnahme hätte signifikante Auswirkung auf:			keine bessere Umweltoption
						Morphologie	Durchgängigkeit	Stau	Schwall	Restwasser	Wasserkraft	Hochwasserschutz	Infrastruktur	
307100000	Tir	Ache	1,66	3,99	4		x			x	x			x
302320000	Tir	Ache	3,99	11,82	4		x			x	x			x
301940001	Tir	Ache	11,82	16,89	3	x	x			x	x			x
101850006	Vbg	Alfenz	3,50	15,50	3		x		x		x			x
101880000	Vbg	Alfenz	15,50	17,00	3		x		x		x			x
100240002	Vbg	Alfenz	17,00	20,50	3						x			x
100910000	Vbg	Alter Rhein	0,00	11,02	5	x						x	x	x
302970000	Sbg	Alterbach	0,00	5,00	4	x	x			x		x	x	x
305610000	Sbg	Altglan	2,50	3,81	4	x						x	x	x
100210007	Vbg	Alvier	0,00	1,50	4	x	x				x	x	x	x
300610000	Sbg	Amerbach	7,36	11,40	4					x	x			x
302320005	Tir	Ampelsbach	0,00	3,00	3		x			x	x			x
901660009	Ktn	Angerbach [Valentinbach]	0,00	4,00	3		x	x				x		x
301750001	Tir	Archbach	0,00	7,26	5	x	x		x	x	x			x
900790017	Ktn	Astenbach	0,00	1,00	4	x	x					x	x	x
1004820000	Bgl	Auwiesenbach	0,00	6,17	4		x					x	x	x
302190001	Sbg	Bluehnbach	0,00	0,70	4	x	x			x		x	x	x
100720002	Vbg	Bolgenach	0,00	7,00	4		x	x			x			x

Source NGPV (2009, att. 1: 1)

individual states. The European Union did not itself ask for scientific data or research but it requested from the member states a professional national river basin management plan that could not be produced without research. The European Union formulated specific criteria the management plans had to fulfil and forced the Austrian Federal Ministry of Agriculture, Forestry, Environment and Water Management to look for scientific information that would meet the obligation.

As a consequence of the pressure by the European Union as an external ally, the Austrian Federal Ministry became a strong internal ally of science. The ministry demanded scientific research and results to solve its task of formulating a sound regulation in line with the European Guideline. The ministry chose scientists and scientific programmes in order to conduct research on specific questions. It communicated closely with the researchers to learn about the potential of research in this topic in Austria and to select relevant questions in detail. The difficult but urgent problem of how to monitor and evaluate the rivers and how to come up with relevant types as a basis for regulation forced the Ministry to take science seriously. Whereas before the pressure by the European Union the Ministry had ignored the scientific results, now it had even started to examine scientific research from previous years. The ministry took part directly in the research, and therefore it played the role of an internal ally. It put its political power behind selected project results and strongly supported the transfer of science to water policy in Austria.

3.6.6.2 Research

Despite focusing on selected questions relevant for practice, the project met high scientific standards. The methods developed and applied are still accessible on the Internet (www.flusslandschaften.at). They are well described, documented and communicated to the scientific community. The project produced many papers published in journals and books (Muhar et al. 2003, 2004), including international scientific journals (Muhar et al. 2005). The project was scientifically innovative in collecting new data about rivers in Austria and in designing new methods for a typology of rivers (UBA 2001, p. 227).

The researchers cooperated with national and international scientific partners. For example, a related, similar project was carried out together with the German Federal Agency for Nature Conservation and with the Bavarian Academy for Nature Conservation and Landscape Management (ANL). The focus was the trans-border implementation of the Guideline for river basin management in Austria and Germany. This project is also documented in scientific publications (Fuchs et al. 2010).

Within the topic of river basin management, the research done by individual projects offered the most suitable information for political actors. The Ministry picked the results from the scientific information that were of high value for the political tasks. All results were scientifically sound, therefore the Ministry's selection did not diminish the scientific quality of the knowledge transferred and used in politics. Disciplined science and methods guarantee high scientific quality, whereas a synthesis by "meta"-science would not be able to meet such a high scientific standard. Such a synthesis was not needed by the Ministry, which had no problems to put the different scientific pieces together guided by the political aims.

3.6.6.3 Integration

Within the project, the scientists had a good idea of the potential demand for their data and for the designed typology. In general, they were orientated toward a sustainable use of rivers and lakes. This public goal provided an umbrella for the research questions. Guided by the aim of sustainable management, the results were able to be linked together in the end.

Even more important was the orientation toward the specific demand by the European Union and the Austrian Federal Ministry of Agriculture, Forestry, Environment and Water Management. The close communication with the ministry enabled the scientists to understand the needs for information well. Due to the requirements for implementing the EU Water Framework Directive in Austria, the task the Ministry had to fulfil was quite clear. The scientists recognised the demand and the mutual trust between them and the Ministry supported the selection of relevant research questions and results. The design of the typology of rivers was to be based on scientific findings in river ecology and on the different uses of rivers, which drive management in practice. Good integration meant that the scientists were well aware of such specific political demands.

The clear and public demands formulated by the European Directive made visible the political actors' need for the scientists. Such information was of major importance if the scientists were willing to specify the particular research questions that might produce a maximum number of relevant results.

The results were presented in a language that best suited the Ministry, which was the most important user. The researchers had this target group in mind when they designed the typology and made tables describing rivers that were close to the documents the Ministry produced. Additionally, an Internet source was designed and produced for the citizen and expert target groups. This link fostered the knowledge transfer by contributing to the participatory processes the Ministry had to fulfil in accordance with the European Union's requisites.

The clear demand by the European Union and the Austrian Federal Ministry of Agriculture, Forestry, Environment and Water Management was met by flexible scientists who made use of the established good communication channels in order to select relevant research questions and results. These were the success factors contributing to an excellent integration of scientific potential with political demands.

3.6.6.4 Utilization

The project produced mainly specialised products for the direct use by the Ministry. The reports were delivered on time and in a format of simple use for the Ministry. The reports considered the professional standards of water policy in Austria and were easy to relate to the professional discourse on these policies. In addition, the scientific arguments were made in awareness of the legal framework of water policy in Austria. The final compatibility with the law was achieved by the Ministry but the scientific reports provided a good basis for the specific data needed by lawmakers.

By means of the internet platform the project provided an important element of environmental governance to the participation. This was not the major aim of the project, but establishing a transparency of methods and results will be an achievement that has positive political impact in the long run. Finally, the project took part in the scientific discourse by means of national and international publications.

3.6.7 *Knowledge Transfer Success Factor List*

The following figure (Fig. 3.24) shows that all the main criteria for RIU were fulfilled in this process. If a particular criterion is of special relevance, it is marked "+".

EU Water Framework Directive
Areas of activity for knowledge transfer (RIU Model)

1	Research	
	Assessing current scientific information	✓ +
	Compliance to the procedures of good scientific practice	✓
	Cooperation with other scientific institutions and projects	✓
	Independent meaningfulness of scientific findings	✓ +

2	Integration	
	Orientation toward public goals (like sustainability, …)	✓ +
	Relevance in regard to political processes	✓ +
	Relevance in regard to allies (external allies)	✓ +
	Target-group oriented intermediation	✓

3	Utilization	
	Contribution to democracy	✓
	Contribution to the rule of law (constitutional legality)	✓ +
	Contribution to "good governance"	✓
	Appropriate solutions to problems	✓ +
	Participation in the scientific discourse	✓

Fig. 3.24 Success factor list for the river landscapes research project

3.7 Research in the Neusiedler See—Seewinkel National Park Region

3.7.1 Project Fact Sheet

- **Project**

  Various research projects in the Neusiedler See – Seewinkel National Park, which crosses the Austro-Hungarian border

- **Type of project**

  Research project

- **Participating research institutions**

  - Neusiedler See – Seewinkel National Park
  - Illmitz Biological Station
  - Various external research institutions

- **Location**

  Neusiedler See – Seewinkel region, State of Burgenland

- **Central question**

  How can the duties performed by the National Park, regulated by law, be organised on a scientific basis?

- **Goals**

  - Provision of the scientific basis for the National Park's sustainable performance of its duties, for example, in the protection of biodiversity in the National Park management plans
  - Long-term ecological monitoring

- **Time frame**

  Since 1993

- **Financing**

  - Austrian Federal Ministry of Agriculture, Forestry, Environment and Water Management (*Österreichisches Bundesministerium für Land- und Forstwirtschaft, Umwelt und Wasserwirtschaft*)
  - State of Burgenland
  - External financing for individual research projects from different sources

3.7.2 Dock Research onto Long-Existing Processes!

The Neusiedler See—Seewinkel National Park (Fig. 3.25) was established in 1993. Its particularity is that, at the time of its establishment it was the only "steppe national park in central Europe" (Ehrenhöfler 1993, p. 2, own translation) and that, in addition, it represents a national park that crosses a border into Hungary. The national park emerged from efforts over decades to protect the Neusiedler See (Lake Neusiedl) area, which had already begun with the foundation of the Austrian League for Nature Conservation, in 1913 (ÖNB 1993, p. 6). After a mass protest against the construction of a pier in the 1970s, in which almost 200,000 people expressed their rejection of this construction project with their signatures, the idea of a national park, which had already been discussed, grew further. At the end of the 1980s, a commission of experts was appointed by Austria and Hungary for this purpose, before the state government of Burgenland began concrete preparations for the establishment of the National Park. Ultimately, the State Parliament of Burgenland adopted the law for the establishment of the National Park, in 1992 (ÖNB 1993, p. 12, 22).

We can therefore ascertain that the establishment of the Neusiedler See—Seewinkel National Park is to do with the result of a process that was long and ongoing, that slowly gained strength and that was supported by many actors.

The National Park Law regulates fundamentally the duties of the National Park. It states:

"(3) The Neusiedler See—Seewinkel National Park Society must ensure the protection of characteristic fauna and flora in the protection zones in accordance with the Management Plan (Paragraph 4), including their habitats and, where needed, existing objects of historical significance and historical or characteristic landscape elements. It must carry out long-term scientific research, ongoing monitoring and preservation of evidence in the protection zones" (NPG 1992, §7). This means that the Neusiedler See—Seewinkel National Park—that has now existed for 20 years—must also carry out research duties, closely linked to its tasks of nature protection and species preservation. The research results "make up the basis for management plans" (Nationalparkverwaltung 2011, p. 28, own translation). Through the duties of the National Park and the need to make management plans for the zone, points of contact have been created here between the activity areas that are important to us: research, integration and utilization. The case study "Neusiedler See—Seewinkel National Park" makes clear that the transfer of the knowledge needed in sustainability research that aims for the long term can be best brought about when research projects "dock onto" solidly institutionalised processes that have existed for a long time. The National Park, which has existed for more than 20 years, and with a historical background now going back 100 years, is a prime example of this.

Given that sustainability research needs scientific information on long-term processes and on the interaction between nature and society, it makes sense to implement numerous zonal research projects in the very same region that build upon each other in the course of years and decades. This is because a truly

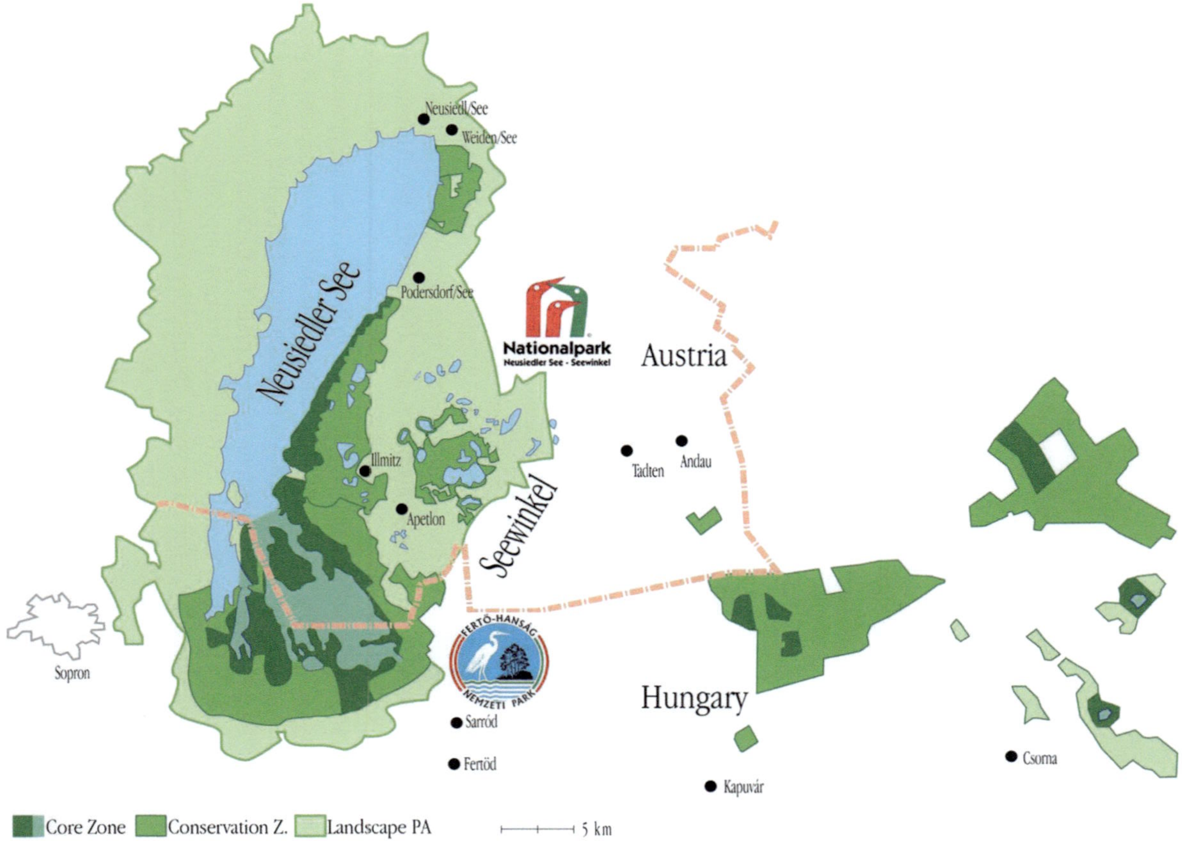

Fig. 3.25 The Neusiedler See—Seewinkel National Park. *Source* Neusiedler See—Seewinkel National Park

sustainable effect upon nature can be expected only from long-term processes. In the same way, it takes decades for societal actors to modify their conduct. For this reason, the link to existing political programmes and funding into which tax money is already brought is important. For long-term issues this is more promising of success than carrying out projects that are isolated and have a short-term time frame. Below we explain in greater depth this success factor for knowledge transfer from the environmental sciences on the basis of examples from the Neusiedler See—Seewinkel National Park.

3.7.3 The Neusiedler See—Seewinkel Region

The special feature of this area is the high proportion of bodies of water and the very high diversity of species. The Lake Neusiedl area is the westernmost steppe lake in Europe, the basin of which covers some 320 km^2, and its reed belt constitutes some 180 km^2 (Nationalpark 2009, p. 5). After that of the Danube delta, the Lake Neusiedl reed belt is the most important in Europe. The region is one of the warmest and also one of the driest landscapes in Austria (Nationalpark 2009, p. 5).

Why is this natural region so important, even internationally? More than 300 bird species use the area as a place for resting and feeding (Nationalpark 2009, p. 5). Of these, 150 species also breed in the Lake Neusiedl area (Nationalpark 2009, p. 5). In addition to this important function as a habitat for numerous species, the Lake Neusiedl and its neighbouring communities are also a popular leisure and recreation area: Particularly for the Viennese, the area represents an important region for local recreation, due to its accessibility (approximately 1 h away by car from Vienna), and in addition, many tourists visit the area to partake in water sports, hike, or bike, or also as nature enthusiasts who participate in the educational and touristic activities on offer by the Neusiedler See—Seewinkel National Park itself. Especially in the summer, the region can be described as being utilized by mass tourism.

In addition to the tourism, which proves itself positive for the lodging businesses and gastronomy, agriculture has an important significance for the region. In particular, viticulture represents an important income source for farmers—typical for the State of Burgenland—and fishing has a certain role as an economic activity for Lake Neusiedl. Additionally, the National Park is a precondition for touristic offerings that have a positive effect on the regional development, for example, even for increasingly specialised offers, like bird watching (Fig. 3.26).

3.7.4 Research Focus of the National Park

We met with a few actors of the Neusiedler See—Seewinkel National Park at the Park's administrative office, in order to learn more about the research carried out in

Fig. 3.26 A lapwing chases a spoonbill at the Neusiedler See—Seewinkel National Park. *Source* Alexander Schneider, Archiv NP Neusiedler See—Seewinkel (*Photo*)

this region. It became fundamentally clear that researchers from the National Park itself carry out only a small portion of the research here, and external researchers conduct a significant portion (Prof. Herzig, Neusiedler See—Seewinkel National Park, personal communication, July 13, 2010). In particular, monitoring is part of the research-related duties of the National Park. For this purpose, the employees of the park or external researchers carry out the research work needed. Monitoring of vegetation ecology and fauna grazing, monitoring of amphibians, management and monitoring of wildlife, monitoring of fish ecology, and ornithological monitoring are particularly important (Nationalparkverwaltung 2011; Nationalpark 2008) (Fig. 3.27).

For the monitoring of grazing, among other things the task consists of preserving and controlling mowing times and good landscape preservation, by farmers, for example, so that agricultural activities do not cause any disturbance of the biodiversity in the area. Similarly, management and monitoring of wildlife consists of recording and controlling the wildlife stock relevant for hunting (Nationalpark 2008, p. 172). The growing roe deer stock led to increasing targets in the governmental hunting quotas (Nationalpark 2008, p. 174). The monitoring of ecological fisheries consists of recording the fish stock in the Lake Neusiedl, as an important biological indicator for the entire ecosystem of the lake (Nationalparkverwaltung 2011, p. 29).

The focus when monitoring the fish stock is to bring the economic significance of eel aquaculture and the marketing of the eels in harmony with the goals of the National Park. Since the eel is not a native species, but an exotic one that was released in the 1950s in the Lake Neusiedl, the National Park's goal of protecting the stock of *local* native species is in conflict with eel fishing. In addition, it was

Fig. 3.27 Sign along a nature path (named after the well-known wildlife biologist and national park activist, Prof. Antal Festetics). *Source* MichaelBöcher (*Photo*)

difficult and expensive for the fishers to acquire the eel larvae necessary for their business activity, because the eel is considered an endangered species. The larvae are necessary because the eels in the Lake Neusiedl cannot reproduce on their own. This is because the Lake Neusiedl does not have a drain, and eels are migratory fish that must move between fresh and salt waters and that therefore need access to other waters in order to spawn naturally. The fishers have accepted the limitations and have now received compensation payments for this. In this case, a mixture of a scientific argument (biodiversity: eels, a non-native species, are not in coherence with the goals and concept of the National Park) and an economic argument (the eel stock is increasingly expensive) led to change.

According to the National Park director, Prof. Herzig, the eel exports diminished from 160 tons per year in the past to only 6 tons per year (personal communication, July 13, 2010). This changed economy within the fishing industry is made clear in the following statement by the local fisheries association: "Since the Lake Neusiedl is a National Park and must be administered according to international guidelines, eels should not be replaced, since the species is not native here" (Fischereiverband Neusiedler See footnote[5]). The idea of the Burgenland Fisheries Association is for the

[5]www.fischereiverband-neusiedlersee.at/index.php?id=3, accessed on 27 October 2013, own translation.

Fig. 3.28 Bird watching in the Neusiedler See—Seewinkel region. *Source* Archiv NP Neusiedler See—Seewinkel (*Photo*)

zander, northern pike, common carp and wels catfish to replace the eel as the main economically important fish in the lake (Nationalpark 2006, p. 143) (Fig. 3.28).

Ornithological monitoring constitutes another focal point. The native bird species in the National Park area have great importance for the National Park, because the diversity of bird species living there increases the attractiveness for visitors: For example, the first international bird-watching fair in the region (1st Pannonian Bird Experience) took place in April 2010, which led to a high number of visitors. Since many of the existing bird species represent rarities, the special task of the National Park is to conduct elaborate ornithological research. For example, the great bittern is located acoustically, and the stock of herons and spoonbills is monitored using aeroplanes. Given that it is precisely the bird watchers who visit the National Park regularly, the bird fauna is an important factor for regional development orientated toward nature conservation: Safeguarding the stocks of rare bird species entices visitors, which in turn has positive effects for business sectors like gastronomy and lodging. In this way, bird watching now represents an important economic factor for touristic businesses at the Lake Neusiedl and also for the optical industry, which, accordingly, markets high-quality optical instruments (e.g., binoculars) (Geschnatter 1/2010: 1) (Fig. 3.29).

Preis: € 1.–
Abo-Preis: € 10.–

NATIONALPARK FRÜHLINGS

GESCHNATTER

Österreichs neunundsiebzigste Nationalparkzeitung 7142 Illmitz, Hauswiese, ✆ 02175/3442, info@nationalpark-neusiedlersee-seewinkel.at • www.nationalpark-neusiedlersee-seewinkel.at Nr. 1 / April 2013

Die Wege des Wassers

Warum es im Feuchtgebiet manchmal sehr trocken ist

Auf die fünf Wintermonate November bis März entfielen von 2011 auf 2012 rund 80 mm Nie-

Fig. 3.29 The National Park journal, Geschnatter ("chatter"). *Source* Geschnatter, 1 April 2013

The different research and yearly reports of the National Park show a series of external research projects that were commissioned by the National Park Society and were carried out in cooperation with different ministries and research institutions. The research report for 2008 documents 14 such projects carried out within the time span of the report, 2007/2008.

The projects illustrate very well the science-based implementation of the focus points mentioned above: monitoring of grazing, of fish ecology, ornithological monitoring and management, and wildlife monitoring. Furthermore, they show the orientation of the research towards the goals of the National Park that are specified in the National Park Law. The different project goals and results found in the reports also make this clear. In this way, in respect of the monitoring of grazing, the following goals are mentioned, among others: "perpetuation and increase of biodiversity in the National Park region" (Nationalpark 2008, p. 153, own translation) and also that "the grazing management […] should be oriented towards the National Park's technical nature conservation goals" (Nationalpark 2008, p. 153, own translation). The following is mentioned as a relevant result: "The spread of reeds along the edges of the saline lakes, which are especially valuable ecologically, was able to be stopped to a great extent by grazing, and areas that were invaded by reeds extensively could be returned to saltmarsh grasses. However, the effects of grazing on the reeds became apparent only after a few years, as a marked decrease in cover values; preparatory mowing accelerated this effect. This reduction improved the habitat conditions for rare plant species of the saltmarsh grasses and the swordleaf rush communities as well as for a series of bird species (shorebirds and other aquatic birds) that depend on this habitat" (Nationalpark 2008, p. 154, own translation).

3.7.5 *The Effects*

Our examples of the research tasks of the Neusiedler See—Seewinkel National Park provide good evidence for the notion that it is sensible to conduct research over long periods of time within one area for purposes of sustainability research. This is because the research conducted here, for example, regarding long-term changes in the vegetation and the interaction between the environment and human influence, cannot be carried out reasonably within projects that have time limitations. On the contrary, research topics at the National Park must be carried out in the long term and permanently. The research projects build on one another. The research resources employed allow the expectation of good results, because they always start from what already exists, and they deliver a synergistic contribution to National Park research as a whole, in addition to highly relevant data.

Being able to perform the monitoring tasks at a high-quality level at all requires the combination of National Park research and external research projects. This is because the resources of the National Park itself do not suffice to ensure the research needed for the professional, high quality, nature-conservation specialist management of national park activities, as the former Research Director of the National Park and Director of the Biological Station in Illmitz, Prof. Herzig, remarked (personal communication, July 13, 2010). Through the governmental financing of external research projects that relate to the same area, other important knowledge emerges, in respect of sustainability goals.

The effects of the research projects presented as examples consist of their generating the imperatively necessary knowledge for the operation of the National Park, and their providing important solutions for regional land users.

3.7.6 *Factors for the Success of Knowledge Transfer*

3.7.6.1 Transfer Through Internal and External Allies

At the National Park Neusiedler See—Seewinkel, internal allies who are participants in the research process themselves look after the support for scientific research and the transfer of scientific knowledge on-site. Among these are the actors in the National Park administration, insofar as they conduct research work independently. Then there are also the cooperation partners of external research projects. They fulfil important integration tasks by implementing the management plans together with the regional actors (land users, citizens) and by orientating the research questions toward the requirements of the actors.

Fishery can be named again as an example. If the management plans for the fish stock at the Lake Neusiedl make provisions for a progressive thinning of eel populations, implementation arises only after the relevant business actors support these science-based results as external allies because of the results' consistency with the actors' own interests. The task of integration is then to develop alternatives for fishery that support the goals of the National Park whilst simultaneously serving the interests of fishers. The change in fish numbers described is evidence for successful integration in the form of the orientation of research towards the development of innovative solutions for specific interests, which then leads to a corresponding practical effect, the result being a positive supply of species diversity of professional nature conservation standards in the National Park and the safeguarding of commercial yields for fishery.

One of the most important external allies is the Austrian Federal Government, which, together with the State of Burgenland, initiated the establishment of the National Park in the 1980s. These external allies provide most of the core funding for the National Park, at 2.5 million euros each (2011), or about 87 % of funding together (figures according to Nationalparkverwaltung 2011, p. 6).

Moreover, the state legislation (Burgenland State Parliament) is an important external ally, because it designated the National Park through the adoption of the National Park Law in 1992, in a democratic decision, and because it standardised the research duty bindingly. Furthermore, the National Park Law documents a broad political consensus regarding the National Park, legitimises it, and strengthens its permanence. The long-term support from the external allies mentioned make possible the synergies that we describe, between the long-term development of a reserve and externally sponsored environmental science and sustainability-related research projects. Within the Austrian Federal Government, the special role of an external ally still corresponds to the Austrian Federal Ministry of Agriculture, Forestry, Environment and Water Management, because it has financed the monitoring programmes and research since 1994 (Nationalparkverwaltung 2011, p. 28). In addition, there are external allies, for example, other, different Austrian ministries or research sponsoring institutions that support National Park-related research projects of a limited duration. These contribute important financial means that support or complement the permanent research tasks to be performed in the National Park.

3.7.6.2 Research

According to our RIU model of successful knowledge transfer, the relevant research must be of good quality and must cover the most current stages of knowledge. In the case of the projects in the Neusiedler See—Seewinkel National Park, the scholarly quality can be determined by whether project results are utilized scientifically, leading up to articles in scientific journals, for example, or by the degree to which scientific cooperation takes place. The available research reports provide information on this. Evidently, the National Park administration itself participates in the research (through Prof. Herzig in fish ecology monitoring, among others) and it cooperates with numerous research institutions, like the University of Vienna (the Dept. of Nature Conservation Biology, and others), the University of Natural Resources and Life Sciences (BOKU), Vienna (Institute for Wildlife Biology and Game Management), the Vienna Museum of Natural History, and diverse private research institutions. Likewise, the Illmitz Biological Station also participates. In addition, there is research cooperation with international partners, for example, with the Institute of Hydrobiology of the Academy of Sciences of the Czech Republic, in the case of fish ecology monitoring.

The research reports document the multitude of publications originating in these projects, both in German and in English, for example *Abhandlungen der Zoologisch-Botanischen Gesellschaft in Österreich* (Transactions of the Royal Zoological-Botanical Society in Vienna) (Waitzbauer et al. 2008), as a collection of results, or articles in international journals like *Bird Study* (Nemeth and Schuster 2005) or *Ecological Modelling* (Nemeth et al. 2005). This, in turn, shows that the research conducted initially for the purposes of the National Park and its management planning needs is relevant beyond this natural region, and that it is also interesting for specific specialized communities. It also shows that the research is of good quality and up-to-date on the most current knowledge. The numerous instances of cooperation (described here briefly) with external Austrian and international scientific institutions are also evidence of the efforts to perform the various monitoring tasks at the highest level of research quality. In this way, knowledge is produced that stimulates the National Park and that leads to technically secure management plans, which are a requirement to fulfil the tasks of the National Park and to safeguard biodiversity. In addition, the knowledge stimulates the expert community beyond the natural region around the Lake Neusiedl.

3.7.6.3 Integration

The most important integration benefit of the research activities at the Neusiedler See—National Park consists of the preparation of a scientific basis for the management plans. These are needed in order for the National Park to carry out its legally-defined duties in nature conservation and in its contributions to the preservation of biodiversity. Here, scientific monitoring results and management requirements are integrated.

Tied in together with this are the relevant solutions for the regional land users. The monitoring activities deliver recommendations and obligations for land users, in order to enable commercial activities in conjunction with the conservation goals of the National Park. Regional development also profits from the result.

However, integration is carried out not only for land users. It also takes place with regard to the local visitors to the National Park and tourists from beyond. The "1st Pannonian Bird Experience", mentioned above, served as a typical integration activity for the information of interested bird watchers. Numerous offers in environmental education are made available, and there is a National Park Centre. Guided excursions convey science-based information to those interested. Every year there is an extensive National Park programme on offer that provides knowledge about the Neusiedler See—Seewinkel area via numerous events and activities, specifically designed for different visitor and age groups. Here, integration at the National Park exists in the fact that such offers are conceived on a scientific basis and then address target groups from the outside directly, in the form of diverse utilization products, for example, excursions for tourists, support for bird watchers, and publications in the National Park's publication, *Geschnatter* (*chatter*).

3.7.6.4 Utilization

According to our model, the generation of utilization products is necessary for a successful knowledge transfer process. At this point, participating scientists no longer have access, because the use of the utilization products is the hand of practical actors. In the National Park research projects described, the results influence the management plans (utilization products) directly, and they are submitted to the ordering customer, the National Park administration, in the form of reports. In addition, the media convey utilization products to different target groups. In this way, this example also reflects typical knowledge transfer processes between research, integration and utilization. Here, there are numerous different products in the utilization, for different target groups: These are, for example, recommendations and obligations for different land users with regard to the measures they shall carry out, and regulations to be observed that are the result of the monitoring activities and the stipulations of the National Park Law. Furthermore, there are utilization products for the most varied visitors: bird watchers, schoolchildren, and tourists, as evidenced by the cover pictures of the relevant programme brochures, as utilization products (see Fig. 3.30).

The National Park journal, *Geschnatter*, published four times a year, reaches the target group "public". It contains information about all activities at the National Park, in commonly understood terms, and it also takes up and explains, on a scientific basis, topics that are controversial in the population, for example, the low water levels in the lake.

In the years 2007/2008 the following projects were carried out:
Monitoring of vegetation ecology and fauna grazing in the Neusiedler See – Seewinkel National Park
Monitoring of fish ecology in the Lake Neusiedl
Ornithological monitoring: monitoring herons and spoonbills of the Nuesiedler See area in the years 2007 and 2008
Ornithological monitoring: brood and migratory stocks of waterfowl in the saline lakes of the Seewinkel in the years 2007 and 2008
Ornithological monitoring: brood stock and brood success of the pied avocet in the Seewinkel (2007, 2008)
Ornithological monitoring: the black-tailed godwit of the Seewinkel (2007, 2008)
Ornithological monitoring: the brood stock of the Kentish plover in the Seewinkel (2007, 2008)
Ornithological monitoring: the brood stock of the common tern in the Neusiedler See area (2007, 2008)
Ornithological monitoring: the brood stock of the common stilt and the Mediterranean gull in the Neusiedler See – Seewinkel National Park (2007, 2008)
Ornithological monitoring: monitoring the hoopoe in the Illmitz-Hölle conservation zone (2007,2008)
Ornithological monitoring: monitoring the great bittern and the great reed warbler in the reed belt of the Lake Neusiedl
(2007, 2008)
Ornithological monitoring: goose stock of the genera *Anser* and *Branta*: migration and winter 2006/2007 as well as 2007/2008 in the Neusiedler See area
The great bustard (*Otis tarda*) in the Waasen-Hanság conservation zone
Monitoring and management of wildlife in the Neusiedler See – Seewinkel National Park

Fig. 3.30 Excerpt from the National Park Research Report, 2008 (own translation). *Source* Nationalpark (2008, p. 152)

Utilization products for the "scientific community" result from the scientific activities, for example, the articles in scientific journals, mentioned above, scientific books, reports, or also events, like professional conferences. These scientific utilization products are documented in the research reports, among other places (Fig. 3.31).

At the Neusiedler See—Seewinkel National Park the utilization of the projects corresponds, for practice, to criteria like appropriateness, democracy and the rule of law: The basis for all National Park activities and the research needed for them is a regional parliament rule which was the result of a democratic process (activities of

Fig. 3.31 Utilization products of the National Park, for different target groups. *Source* Nationalpark Neusiedler See—Seewinkel (2012a, b, 2013)

environmental organisations, agents acting on their convictions, and other National Park activists) that took a decade. The National Park Law provides a formal democratic legitimation of the National Park research (democratic criteria) and considers long-term ecological and social goals, like sustainability and preservation of biodiversity. The utilization products meet the demands of appropriateness—they produce solutions that are applicable to existing problems (requirements of the National Park management, needs of different land users). The solutions developed relate to current political problems, since the establishment of a national park decided through policy generates problems constantly in practice, which are alleviated and solved by means of scientific, innovative solutions. In this way, the utilization products observed in this example also meet the quality criteria in terms of appropriateness of solutions.

3.7.7 Knowledge Transfer Success Factor List

The following figure (Fig. 3.32) shows that all the main criteria for RIU were fulfilled in this process. If a particular criterion is of special relevance, it is marked "+".

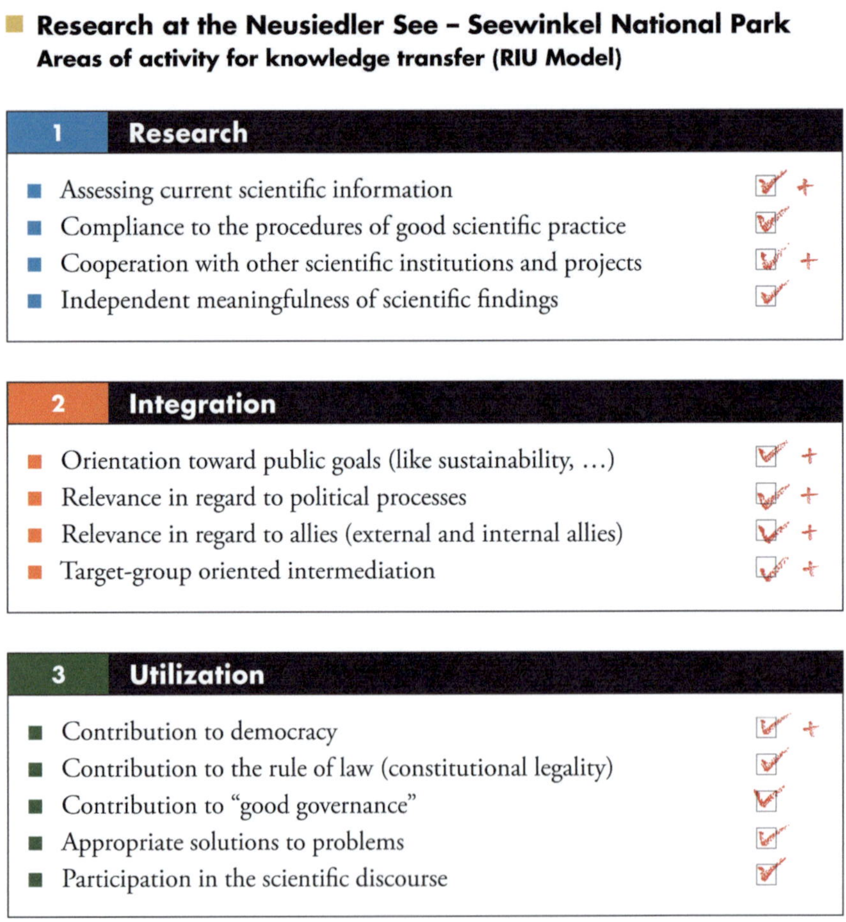

Fig. 3.32 Success factor list for the Neusiedler See—Seewinkel National Park research

References

Begusch-Pfefferkorn, K. (2006). KLF-Wissensbilanz - Leistungsdarstellung des inter- und transdisziplinären Programms Kulturlandschaftsforschung. In B. Glaeser (Ed.), *Fachübergreifende Nachhaltigkeitsforschung* (pp. 119–139). München: Oekom.

Böcher, M. (2008). Regional governance and rural development in Germany: the implementation of LEADER+. *sociologia ruralis, 48*(4), 372–388.

Böcher, M., & Krott, M. (2012). Nachhaltigkeitsforschung in Österreich. *GAIA-Ecological Perspectives for Science and Society, 21*(2), 155–157.

Böcher, M., & Krott, M. (2014). *Mit Wissen bewegen! Erfolgsfaktoren für Wissenstransfer in den Umweltwissenschaften*. München: Oekom.

Böcher, M., & Töller, A. E. (2012). *Umweltpolitik in Deutschland: eine politikfeldanalytische Einführung* (Vol. 50). Wiesbaden: VS Springer-Verlag.

Brand, K.-W. (2000). Nachhaltigkeitsforschung - Besonderheiten, Probleme und Erfordernisse eines neuen Forschungstypus. In K.-W. Brand (Ed.), *Nachhaltige Entwicklung und Transdisziplinarität* (pp. 9–28). Berlin: Analytica.

Bundesministerium für Bildung, Wissenschaft und Kultur (2005). KLF-Resultate. Empfehlungen, Indikatoren, neues Wissen aus der österreichischen Kulturlandschaftsforschung CD-Rom (Forschungsprogramm Kulturlandschaft). Wien: bm:bwk.

Bundesministerium für Land- und Forstwirtschaft, Umwelt und Wasserwirtschaft (2006). Eine Leitlinie für unser Wasser. Die Europäische Wasserrahmenrichtlinie. Wien: BMLFUW.

Bundesministerium für Land- und Forstwirtschaft, Umwelt und Wasserwirtschaft (2007). Wasser in Österreich. Zahlen und Fakten. Wien: BMLFUW.

Bundesministerium für Land- und Forstwirtschaft, Umwelt und Wasserwirtschaft (2011). Wasserzeichen. Wien: BMLFUW.

Bundesministerium für Wissenschaft, Forschung und Kultur (1995). Forschungskonzept 1995 Kulturlandschaftsforschung. Wien: BMWFK.

Bundesministerium für Wissenschaft und Forschung. (2007). *Zweite Ausschreibung des Förderprogramms proVISION: Vorsorge für Natur und Gesellschaft*. Wien: BMWF.

Deutscher Verband für Landschaftspflege. (2010). Kleine Fließgewässer kooperativ entwickeln. Erfolgsmodelle für die Umsetzung der Wasserrahmenrichtlinie (Vol. 17, DVL-Schriftenreihe Landschaft als Lebensraum). Ansbach: DVL.

Ehrenhöfler, E. (1993). Eine Vision wird Wirklichkeit. In Ö. N. (ÖNB) (Ed.), Schritt für Schritt zum Nationalpark Neusiedler See / Seewinkel (pp. 2).

European Commission. (2001). *Economy-wide material flow accounts and derived indicators: a methodological guide*. Luxembourg: Office for Official Publications of the European Commission.

European Environment Agency. (2010). *The European Environment: State and Outlook 2010: Synthesis*. Copenhagen: European Environment Agency.

EUROSTAT (2009). Economy Wide Material Flow Accounts: Compilation Guidelines for reporting to the 2009 Eurostat questionnaire.

Federal Ministry of Agriculture, Forestry, Environment and Water Management (Bundesministerium für Land- und Forstwirtschaft, Umwelt und Wasserwirtschaft) BMLFUW. (2010). *Die Umweltgesamtrechnungen in Österreich*. Wien: BMLFUW.

Federal Ministry of Agriculture, Forestry, Environment and Water Management. (2013). *Sustainability Barometer 2013*. Federal Ministry of Agriculture, Forestry, Environment and Water Management: Headline Indicators. Vienna.

Federal Ministry of Agriculture, Forestry, Environment and Water Management (Bundesministerium für Land- und Forstwirtschaft, Umwelt und Wasserwirtschaft), & Bundesministerium für Wirtschaft Familie und Jugend. (2011). *Ressourcennutzung in Österreich - Bericht 2011*. Wien: BMLFUW/BMWFJ.

Fink, M. H., Moog, O., & Wimmer, R. (2000). Fließgewässer-Naturräume Österreichs - eine Grundlage zur typologischen Charakteristik österreichischer Fließgewässer (Vol. 128, Monographien des Umweltbundesamtes). Wien: UBA.

Fischer-Kowalski, M., & Haberl, H. (1997). Tons, joules, and money: Modes of production and their sustainability problems. *Society & Natural Resources, 10*(1), 61–85.

Fischer-Kowalski, M., Haberl, H., Payer, H., Steurer, A., Zangerl-Weisz, H., &, W. I. (1991). Verursacherbezogene Umweltindikatoren - Kurzfassung, Endbericht Teil I des Projekts "Verursacherorientierte Umweltindikatoren" des IFF/Ökologie-Institutes. Wien: IFF.

Fischer-Kowalski, M., Bittermann, W., Haberl, H., Weisz, H., & Winiwarter, V. (1999). Versuch einer Synthese: Ein Gruppeninterview. In H. Haberl, W. Bittermann, W. Hüttler, H. Weisz, M. Fischer-Kowalski, H. Schandl, et al. (Eds.), Colonizing Landscapes—Indicators for Sustainable Land Use, Synthesebericht zum Kulturlandschaftsforschungs-Modul IN4, Top-down Planungsindikatoren (pp. 125–149). Wien.

Freyer, B., & Muhar, A. (Eds.). (2006). *Transdisziplinäre Kooperation in der universitären Ausbildung: die Fallstudie "Leben 2014" in der Nationalparkregion Hohe Tauern, Oberpinzgau; ein Projekt im Rahmen des Programms Kulturlandschaftsforschung*. Wien: Facultas.

Fuchs, M., Preis, S., Wirth, V., Binzenhöfer, B., Pröbstl, U., & Pohl, G., et al. (2010). Wasserrahmenrichtlinie und Natura 2000. Gemeinsame Umsetzung in Deutschland und Österreich am Beispiel der Grenzflüsse Salzach und Inn (Vol. 85, Naturschutz und Biologische Vielfalt). Bonn: Bundesamt für Naturschutz.

Garrelts, H., & Krott, M. (2002). Erfolg und Versagen roter Listen-wann ist deren Einsatz ratsam. *Natur und Landschaft, 77*(3), 110–115.

Geschnatter – Zeitung des Nationalparks Neusiedler See-Seewinkel verschiedene Ausgaben.

Glanzer, M., & Muhar, A. (2006). Organisatorischer Rahmen des Projektes „Leben 2014". In B. Freyer & A. Muhar (Eds.), *Transdisziplinäre Kooperation in der universitären Ausbildung: die Fallstudie "Leben 2014" in der Nationalparkregion Hohe Tauern, Oberpinzgau; ein Projekt im Rahmen des Programms Kulturlandschaftsforschung* (pp. 1–11). Wien: Facultas Universitätsverlag.

Glanzer, M., & Schauppenlehner, T. (2006). Umsetzungsprojekte. In B. Freyer, & A. Muhar (Eds.), Transdisziplinäre Kooperation in der universitären Ausbildung. Die Fallstudie "Leben 2014" in der Nationalparkregion Hohe Tauern/Oberpinzgau (pp. 208–233). Wien: Facultas Universitätsverlag.

Glanzer, M., Freyer, B., Muhar, A., Schauppenlehner, T., & Vilsmaier, U. (2005). *Leben 2014: Perspektiven der Regionalentwicklung in der Nationalparkregion Hohe Tauern, Oberpinzgau; Dokumentation der Ergebnisse;[ein Projekt im Rahmen des Programms Kulturlandschaftsforschung]*. Neukirchen/Großvenediger: Verlag Tauriska.

Gutheil-Knopp-Kirchwald, G. (2004). Lohnt sich die Umstellung auf Direktvermarktung für einen Wiener Produzenten? *Land und Raum, 5*(3), 9–16.

Haberl, H. (1997a). Human Appropriation of Net Primary Production as an Environmental Indicator: Implications for Sustainable Development. *Ambio, 26*(3), 143–146.

Haberl, H. (1997b). Räumliche Verteilung der menschlichen Aneignung oberirdischer Nettoprimär-produktion in Österreich. *Verhandlungen der Zoologisch-Botanischen Gesellschaft, 134*, 413–430.

Haberl, H., Bittermann, W., Hüttler, W., Weisz, H., Fischer-Kowalski, M., Schandl, H., et al. (Eds.). (1999). Colonizing Landscapes—Indicators for Sustainable Land Use. Synthesebericht zum Kulturlandschaftsforschungs-Modul IN4, Top-down Planungsindikatoren. Vienna.

Haberl, H., Amann, C., Bittermann, W., Erb, K.-H., Fischer-Kowalski, M., Geissler, S., et al. (2001). *Die Kolonisierung der Landschaft: Indikatoren für nachhaltige Landnutzung (Forschungsschwerpunkt Kulturlandschaft 8)*. Wien: Bundesministerium für Bildung, Wiss. u. Kultur.

Haberl, H., Erb, K. H., Krausmann, F., Gaube, V., Bondeau, A., Plutzar, C., et al. (2007). Quantifying and mapping the human appropriation of net primary production in earth's terrestrial ecosystems. *Proceedings of the National Academy of Sciences, 104*(31), 12942–12947.

Hüttl, R. (2003). *Internationale Evaluierung des Programms Kulturlandschaftsforschung (KLF)*. Wien: Im Auftrag des BMBWK.

Kaindl, A. (2006). Skitourismus droht das Aus. *Salzburger Nachrichten., 12*(9), 2006.

Kromp, B., Hartl, W., & Maurer, L. (2004). Ökologisierung der urbanen Landwirtschaft in Wien: Status Quo-Erhebung und Zukunftsmöglichkeiten. Land und Raum(3), 21–25.

Landauer, M., & Pröbstl, U. (2008). Klimawandel, Skilanglauf undTourismus in Österreich. *Naturschutz und Landschaftsplanung, 40*(10), 336–341.

Mader, H., Steidl, T., & Wimmer, R. (1996). Abflussregime österreichischer Fließgewässer - Beitrag zu einer bundesweiten Fließgewässertypologie (Vol. 82, Monographien des Umweltbundesamts). Wien: UBA.

Maurer, L., Meindl, P., Erhart, E., Forster, A., Hartl, W., Kienegger, M., et al. (2002). Optionen für die Entwicklung von Landwirtschaft und Gartenbau in Wien. Endbericht des Forschungsprojektes im Auftrag des Bundesministeriums für Bildung, Wissenschaft und Kultur und der Magistratsabteilung 22, Umweltschutz der Stadt Wien. Wien.

Meyer-Cech, K., & Seher, W. (2004). Beitrag der Stadtlandwirtschaft zur Sicherung städtischer Freiräume, am Beispiel der Stadt Wien. *Land und Raum, 5*(3), 4–8.

Milota, E., & Petrović, B. (2012). Umweltgesamtrechnungen, Modul Materialflussrechnung (Zeitreihe 1995 bis 2010), Projektbericht. Wien: Statistik Austria (Im Auftrag des Bundesministeriums für Land- und Forstwirtschaft, Umwelt und Wasserwirtschaft).

Moog, O., & Wimmer, R. (1990). Grundlagen zur typologischen Charakteristik österreichischer Fließgewässer. *Wasser und Abwasser, 34*, 55–211.

Muhar, S., Kainz, M., Kaufmann, M., & Schwarz, M. (1996). Ausweisung flußtypspezifisch erhaltener Fließgewässerabschnitte in Österreich - Bundesflüsse lt. §8 WBFG,. Wien: Bundesministerium für Land- und Forstwirtschaft.

Muhar, S., Kainz, M., & Schwarz, M. (1998). *Ausweisung flußtypspezifisch erhaltener Fließgewässerabschnitte in Österreich - Fließgewässer mit einem Einzugsgebiet > 500 km2 ohne Bundesflüsse*. Wien: Bundesministerium für Land- und Forstwirtschaft.

Muhar, S., Poppe, M., Egger, G., Schmutz, S., & Melcher, A. (2004). Flusslandschaften Österreichs. Ausweisung von Flusslandschaftstypen anhand des Naturraums, der Fischfauna und der Auenvegetation (Forschungsprogramm Kulturlandschaft). Wien: bm:bwk.

Muhar, A., Vilsmaier, U., & Freyer, B. (2006). The polarity field concept–A new approach for integrated regional planning and sustainability processes. *GAIA-Ecological Perspectives for Science and Society, 15*(3), 200–205.

Muhar, S., Poppe, M., Schmutz, S., Egger, G., & Melcher, A. (2003). Analyse und Ausweisung naturräumlicher Flusstypen Österreichs. *Österreichische Wasser- und Abfallwirtschaft, 7–8*, 113–121.

Muhar, S., Schmutz, S., Preis, S., Jungwirth, M., & Egger, G. (2005). Integrative river management in the context of the EU Water Framework Directive: bringing together ecological and socioeconomic requirements of river landscape development. In E. Lange & D. Miller (Eds.), *Our shared landcape*. Ascona: Monte Verita.

Municipal Department 58 (Magistratsabteilung 58) Agrarstruktureller Entwicklungsplan für Wien. Abschlussbericht. Wien: MA 58.

Nationalpark Neusiedler See-Seewinkel (Nationalpark 2006) Forschungsbericht 2006. Illmitz.

Nationalpark Neusiedler See-Seewinkel (Nationalpark 2008) Forschungsbericht 2008. Illmitz.

Nationalpark Neusiedler See-Seewinkel (Nationalpark 2009). (2009). Das Wichtigste für Besucher. Illmitz.

Nationalpark Neusiedler See-Seewinkel (Nationalpark 2012). (2012a). Schulprogramme 2012/13. Illmitz.

Nationalpark Neusiedler See-Seewinkel (Nationalpark 2012). (2012b). Programm Jänner 2013 - Feber 2014. Illmitz.

Nationalpark Neusiedler See-Seewinkel (Nationalpark 2013). (2013). Programmheft zur Pannonian Bird Experience 2013. Illmitz.

Nationalparkverwaltung Neusiedler See-Seewinkel. (2011). Arbeitsbericht 2011. Apetlon.

Nemeth, E., & Schuster, A. (2005). Spatial and temporal variation of habitat and prey utilization in the Great White Egret Ardea alba alba at Lake Neusiedl, Austria. *Bird Study, 52*(2), 129–136.

Nemeth, E., Bossew, P., & Plutzar, C. (2005). A distance-dependent estimation of foraging ranges of neighbouring bird colonies. *Ecological Modelling*, *182*(1), 67–73. doi:http://dx.doi.org/10.1016/j.ecolmodel.2004.07.011.

NGPV. (2009). Nationale GewässerbewirtschaftungsplanVO 2009 (NGPV 2009), BGBl. II Nr. 103/2010, ausgegeben am 30. März 2010.

NPG. (1992) Gesetz über den Nationalpark Neusiedler See-Seewinkel Landesgesetzblatt 1993, 9. Stück v. 11. Feber 1993.

Österreichischer Naturschutzbund (ÖNB). (1993). Schritt für Schritt zum Nationalpark Neusiedler See/ Seewinkel (= Natur und Land, 79. Jg. Sonderheft zu 3-4/1993). Salzburg.

Österreichisches Normungsinstitut. (1997). ÖNORM M 6232 Richtlinien für die ökologische Untersuchung und Bewertung von Fließgewässern. Wien.

Paula, M., Smoliner, C., & Tiefenthaler, B. (2004). *Forschung für nachhaltige Entwicklung FORNE - Rahmenstrategie 2004+*. Wien: BMVit, BMBWK, Lebensministerium.

Pesendorfer, D. (2007). *Paradigmenwechsel in der Umweltpolitik. Von den Anfängen der Umwelt- zu einer Nachhaltigkeitspolitik: Modellfall Österreich?*. Wiesbaden: VS Verlag für Sozialwissenschaften.

Pregernig, M., & Böcher, M. (2012). Normative and analytical perspectives on the role of science and expertise in environmental governance. In K. Hogl, E. Kvarda, R. Nordbeck, & M. Pregernig (Eds.), *Environmental Governance: The Challenge of Legitimacy and Effectiveness* (pp. 199–219). Cheltenham: Edward Elgar.

Pröbstl, U. (2007). Klimawandel: Zukunft und Herausforderung für den Tourismus. Ländlicher Raum, 1–13.

Pröbstl, U., & Prutsch, A. (2008a). Endbericht STRATEGE. Strategien zur nachhaltigen Raumentwicklung von Tourismusregionen unter dem Einfluss der globalen Erwärmung am Beispiel der Wintersportregion um Schladming. Wien: Universität für Bodenkultur (BOKU).

Pröbstl, U., & Prutsch, A. (2008b). Klimawandel und Wintertourismus - Bedrohung oder Panikmache. *FdSnow, 26*(32), 35–40.

Professional Association of Austrian Cable Cars (Fachverband der Seilbahnen). (2007). Klimawandel in der öffentlichen Diskussion, Lokale Fakten statt pauschaler Urteile, Presseinformation, Pressekonferenz, 14.11.2007.

Rogall, H. (2008). Ökologische Ökonomie. Eine Einführung (2ed.). Wiesbaden: VS Verlag für Sozialwissenschaften.

Schneidewind, U., & Singer-Brodowski, M. (2013). *Transformative Wissenschaft. Klimawandel im deutschen Wissenschafts- und Hochschulsystem*. Marburg: Metropolis.

Spangenberg, J. H. (1996). Welche Indikatoren braucht eine nachhaltige Entwicklung. In J. Köhn & M. J. Welfens (Eds.), *Neue Ansätze in der Umweltökonomie* (pp. 203–225). Marburg: Metropolis.

Stadtentwicklung Wien (2002). Landwirtschaft in Wien - zwischen Stadtplanung und Ökonomie. Vienna: MA 18.

Stadtentwicklung Wien. (2005). STEP 05 - Stadtentwicklung Wien 2005. Wien: MA 18.

Turnhout, E., Hisschemöller, M., & Eijsackers, H. (2007). Ecological indicators: Between the two fires of science and policy. *Ecological Indicators, 7*(2), 215–228. doi:10.1016/j.ecolind.2005.12.003.

Umweltbundesamt. (2001). 6. Umweltkontrollbericht. Wien: UBA.

Unbehaun, W., Pröbstl, U., & Haider, W. (2008). Trends in winter sport tourism: challenges for the future. *Tourism Review, 63*(1), 36–47. doi:10.1108/16605370810861035.

Vilsmaier, U. (2006). Weiterführung und Projektumsetzung in der Region. In B. Freyer, & A. Muhar (Eds.), Transdisziplinäre Kooperation in der universitären Ausbildung. Die

Fallstudie "Leben 2014" in der Nationalparkregion Hohe Tauern/Oberpinzgau (pp. 278–280). Wien: Facultas.

Waitzbauer, W., Korner, I., & Wrbka, T. (2008). Vegetationsökologisches und faunistisches Beweidungsmonitoring im Nationalpark Neusiedler See-Seewinkel 2000–2006 (Vol. 37, Abhandlungen der Zoologisch-Botanischen Gesellschaft in Österreich): Verlag der Zoologisch-Botanischen Gesellschaft in Österreich.

Weisz, H., Krausmann, F., Erb, K.-H., Schulz, N., & Haberl, H. (1999). Gesellschaftliche Beeinflussung ökosystemarer Energieflüsse und Energiebestände (NPP-Aneignung). In H. Haberl, W. Bittermann, W. Hüttler, H. Weisz, M. Fischer-Kowalski, H. Schandl, et al. (Eds.), Colonizing Landscapes Indicators for Sustainable Land Use. Synthesebericht zum Kulturlandschaftsforschungs-Modul IN4 „Top-down Planungsindikatoren" (pp. 29–54). Wien.

Whitelegg, K. (2002). Identifying and Assessing National Research Activities on Sustainable Development. *Technikfolgenabschätzung, 11*(3/4), 128–133.

Whitelegg, K. (2006). The (re)search for solutions: research programmes for sustainable development. In J.-P. Voß & D. Bauknecht (Eds.), *Reflexive Governance for Sustainable Development* (pp. 273–293). Cheltenham: Edward Elgar.

Whitelegg, K., Weber, M., & Leone, F. (2002). *National research activities and sustainable development (IPTS Technical Report Series, EUR 20389 EN)*. Sevilla: European Commission Joint Research Centre, IPTS - Institute for Prospective Technological Studies.

Wimmer, R., & Moog, O. (1994). Flußordnungszahlen österreichischer Fließgewässer (Vol. 51, Monographien des Umweltbundesamtes). Wien: UBA.

World Tourism Organization. (2003). Climate Change and Tourism—Proceedings. In W. T. Organization (Ed.), 1st International Conference on Climate Change and Tourism, Djerba, Tunesia, World Tourism Organization.

Chapter 4
Theoretical Foundations of RIU

4.1 Towards a Political Science Explanation of Scientific Knowledge Transfer

The starting point of our RIU model of scientific knowledge transfer is a wish to better explain the conditions under which environment-related scientific information can influence political (or practical) decisions. Our view on scientific knowledge transfer is from the perspective of political science. A political science view means that we believe that decisions in society are the result of multiple power relations between multiple actors and that actors try to realise their interests (Krott 2012), or, as Warren puts it, that political practice is orientated toward organising collective action in the context of differing short-term interests and the power relationships that exist in modern society (Warren 1999). Decisions are political, which means that they rely at least partly on normative aspects that cannot be decided scientifically. Scientific information cannot fully guide political decisions; rather, scientific information can inform political actors for science-related matters so as to help make political decisions. We also believe that, to a great extent, political decisions are made without referring to any scientific information: They are based on underlying power relations, political ideologies, traditions, political experience (from the past), or they are even simply made at random.

An important research question is therefore how—even under the conditions described above—scientific information can have an effect on political decision making. In the following subchapters we will present a brief overview of selected approaches on scientific knowledge transfer, guided by an assessment of how these approaches are able to illuminate the conditions under which science can have an effect on political decisions.[1] Following this brief review, we will elaborate further on the theoretical foundations of our RIU model and on why we think that RIU is

[1]See, fore more comprehensive overviews, e.g. Pregernig (2014), Pregernig and Böcher (2012), Weingart (2003), Jasanoff and Wynne (1998).

M. Böcher and M. Krott, *Science Makes the World Go Round*,
DOI 10.1007/978-3-319-34079-1_4

able to deliver an insightful explanation of the conditions under which science is able to become part of political decisions.

4.2 Linear or Technocratic Models

Linear models of scientific knowledge transfer have been labelled as "technocratic" (Hulme 2009, p. 102); they propose that scientific knowledge can be directly applied to politics because it is upstream from political decision making (Beck 2011, p. 298). Science helps to increase the effectiveness and efficiency of political decisions. Scientific knowledge leads to better policies and "evidence-based" policy making (see e.g. Nutley et al. 2000). If policy-relevant scientific facts are available, then this availability automatically leads to the use of science in politics (Hulme 2009, p. 101). The idea behind linear knowledge transfer models is that there is a strict separation between the different spheres of science and politics as well as between scientific facts and political norms. In linear models, scientific knowledge is demanded by political actors and is supplied by the scientific community: this knowledge flows directly from science into politics, in which it is utilized by political decision makers with the aim of producing "rational" solutions (Guston 2001, p. 100). Political decisions are based on scientific information without being altered by the policy process. Technocracy may even replace political decisions due to inevitable scientific facts that predetermine decisions in a way that politics becomes redundant (Schelsky 1965). Science is applied simply because politics needs science in certain cases. Linearity means that if more scientific knowledge is produced then political decisions are also scientifically informed following a proportional relationship: The more science is produced—the more political decisions are shaped by science. The linear model of scientific knowledge transfer is still highly relevant among natural or climate scientists (Durant 2015) who often ask, critically, why political actors do not follow their recommendations if they see empirical deviations from the linear model of scientific knowledge transfer (Beck 2011, p. 298).

In our view, linear models are useful to explain those cases in which science-based information leads to political decisions that nearly fully reflect the respective scientific information. That is sometimes the case if concrete scientific data or methods are necessary to implement policies. The case we studied, on the implementation of the EU water framework directive in Austria (Sect. 3.6) illustrates this: Due to the demand for science-based monitoring of Austrian waters, scientific information played a role in this case, approaching the linear model of scientific knowledge transfer. However, the question remains of whether science-based information is applied solely due to its "scientific" character, or rather because of the interests of politics and an administration that need science-based information to serve their interests in order to implement the new European directive.

Therefore, in our view, the linear model overestimates the possibility of science being integrated into political decisions due to its scientific rationality, and it underestimates the role of power and the interests of different political actors as a more important factor for the application of science. Furthermore, linear models lack explanatory power in cases in which political decisions deviate from pure scientific knowledge for political reasons.

From our political science perspective, a direct, "linear" knowledge transfer from science ("truth") to political practice ("power") seldom appears to be possible. What are the critical points in linear models? As we have observed, many "linear" models of scientific knowledge transfer argue that the availability of scientific research alone leads to a transfer of scientific knowledge into political practice (Grundmann 2009, p. 398). These models argue that policy makers are interested in solving problems (problem-solving bias) (Böcher and Töller 2015) and are therefore actively searching for and using scientific evidence. However, modern public policy analysis has revealed that real-world policy processes are much more non-linear and that public policy is not necessarily aimed at solving problems (Dunn 2015; Böcher and Töller 2012; Kingdon 2003). "It is a commonplace of political science that politics is not primarily about effective problem solution, but about gaining and maintaining power." (Mayntz 1983, p. 333). In politics "substantive rationality ranges relatively low in the order of political decision criteria" (Mayntz 1983, p. 333)

The space for "rational" science-based solutions is much smaller than is theorised, as political actors follow their self-interests, which do not necessarily include the maximal use of science (Braun and Benninghoff 2003). Therefore, the use of scientific expertise is by no means dependent only on scientific quality but is also on its usefulness for various political actors (Miller 2008, p. 148). Another main assumption of the linear model is that there is always an uncontested "best" solution to a policy problem. The reality of science, especially regarding particularly complex problems, like environmental issues, is much more contested within science itself than the linear model argues (Beck 1992). As Parsons points out, it is often the case that "For every expert who says A, there are experts who can say B with equal claim to professional or expert authority" (Parsons 1995, p. 158). The existence of scientific uncertainties is typical for the science system and restricts the possibility of linear knowledge transfer (Bradshaw and Borchers 2000).

The linear model has its strength in explaining scientific knowledge transfer in cases in which political solutions follow scientific expertise clearly. The model lacks explanatory power when analysing political factors that lead to scientific knowledge transfer or political factors that hinder the practical utilization of science in politics. This is because the linear model argues that political decisions are based on science due to scientific rationality. It does not explain cases in which political actors are not interested in problem-solving as such but rather, apply scientific knowledge for reasons other than science itself.

4.3 Different System Models

Some theoretical models of scientific knowledge transfer highlight the natural incompatibilities between science and politics (Miller 2008, p. 148). The reason for these incompatibilities is that both societal systems are based on fundamentally different rationalities (Lehmann and Rieder 2003): Science is an incremental process that is oriented toward finding the truth by using sound scientific standards and methodologies (Ascher et al. 2010, p. 6). This situation implies the existence of scientific controversies (Bocking 2004, p. 4; Bradshaw and Borchers 2000) and long-time research procedures before practical implications are even detected (Ascher et al. 2010, p. 1). Normally, scientific progress needs time and science has all the time it needs to improve existing methods and theories. Pure science is only concerned with improving its own findings and getting closer to the truth without necessarily involving political and/or practical questions. Academic peers within the scientific system judge scientific findings.

In contrast to this, the function of the political system is the production of political decisions. These decisions are based on political rationality, which includes interests, power and political ideologies. In politics, many aspects other than "just" scientific reasoning lead to political decisions. What is also typical of political decisions is that they have to be taken under conditions that include time limitations and uncertainties. This contrasts strongly with science, which seeks to reduce uncertainty by investing time in theory development and producing science-based evidence. These fundamental differences between the system of science and the political system impede scientific knowledge transfer structurally. It was the famous German sociologist, Niklas Luhmann, who argued that different social systems follow their own rationalities. Whereas the science system's specific communication medium is "truth" (Luhmann 1968, p. 148), the specific medium of communication of political systems' is "power" (Luhmann 1981, p. 287). The function of the science system lies in the production of new knowledge through research processes, whereas the function of the political system lies in the production of collectively binding decisions through power processes (Luhmann 1981, p. 287). According to this strand of systems theory, scientific knowledge transfer means the conversion of scientific information into that of another system (in our case, that of politics). However, according to Luhmann, in principle communication is impossible between different systems. Each system is "self-referential" and closed to the other. These problems are structural by nature and are by far more than just problems of communication or simple transformation (Willems et al. 1993, p. 47). If we take these theoretical descriptions of the different systems that are relevant for scientific knowledge transfer seriously, it becomes much clearer why it is so difficult to find utilization of scientific knowledge in political practice. Actors representing either system have structural difficulties in understanding each other, since they act as a part of different system rationalities. Scientific knowledge transfer cannot be improved easily by communication or translation activities, since fundamental structural differences persist even despite the best possible communication efforts.

The value of systems theory approaches is that they explain theoretically the important difference between science and politics that makes it so difficult to simply transform scientific findings into practical utilization. With the help of systems theory it is possible to explain the main reasons for problems in scientific knowledge transfer in a structural, meta-theoretical manner. Such approaches can help us to better explain why politicians cannot fully understand scientists and vice versa, as we have seen in Chap. 1 of this book. By only referring to pure systems theory approaches, however, it becomes difficult to explain cases in which scientific knowledge transfer can be observed, even though politicians do not understand science and scientists do not understand politics.

4.4 Functional Models

As mentioned above, politics is oriented toward organising collective action in the context of differing short-term interests and the power relationships that exist in modern society (Warren 1999). Therefore, political actors often use simple messages (in contrast to scientific arguments) or need scientific results at the exact moment that policy windows are open for a certain political decision (Kingdon 2003). In many cases, political actors must decide immediately and cannot wait for science to produce more information. Miller explains this point in an illuminating manner: "It is generally acknowledged that environmental policies ought to be guided by sound science, but the scientific process and public policymaking are fundamentally incompatible. This does not stop the combatants in the policy wars from exploiting science for their own ends" (Miller 2008, p. 148). Miller draws our attention to the idea that environmental policy should be guided by science (Bocking 2004; Gupta et al. 2013; Haas 2005), but in reality, science is often neglected by politicians simply because the rationalities of politics differ from those of science completely. In most cases, political actors take decisions due to their interests, which can, but need not, include science-based reasoning. Functional models are based on this model of politics, which stems from public choice theory (Mueller 2004; for an application in environmental policy analysis, see Kirchgässner and Schneider 2003), and describe functions of science for political actors: The main idea is that science is not utilized for its epistemic features but—if not completely neglected—it is used only because it can fulfil several *functions* for political actors: Boehmer-Christiansen has described several functions of science for politics (Boehmer-Christiansen 1995):

- Scientific expertise may be a source of authority and, hence, legitimacy for those in power: Politicians use scientific knowledge selectively to legitimize their political decisions, especially if these decisions include unpopular political measures (Boehmer-Christiansen 1995).

- Scientific expertise may act as an instrument of persuasion in debates and negotiations (with the parties tending to select the recommendations that best fit their own interests): Different politicians will use scientific knowledge to argue their interests in political debates and to persuade competing actors to embrace their political interests (Boehmer-Christiansen 1995, p. 197).
- Scientific expertise may function as a mechanism for delaying or avoiding action, or as a substitute for action (because conducting more research buys time and transfers the responsibility to someone else). Politicians often avoid political decision making in high-conflict fields by deferring to the lack of scientific expertise that would be necessary to make a decision. Such actors may also establish a scientific commission or initiate scientific research projects to demonstrate their political action (Boehmer-Christiansen 1995, p. 197).
- Scientific expertise may facilitate a policy change by acting as a scapegoat (because science may be used to allow politicians to change their minds without a loss of reputation or admission of error) (Boehmer-Christiansen 1995, p. 197).

Functional models emphasise the egoistical preferences of rational political actors. Political actors use scientific knowledge simply as an additional resource that will increase their authority or legitimacy. They do not have an interest in scientific rationality or the results of scientific research. The scientific knowledge that they use has no epistemic function; it simply provides an additional way for them to legitimise their interests and to gain power (Radaelli 1995). Competing parties choose only advice that supports their own policy choices and overlook advice that does not (Sarewitz 2004; Boehmer-Christiansen 1995). Decision makers try to use their expertise to legitimate their interests and political programmes and thus support those scientists who deliver the scientific results that conform to their expectations and beliefs (Schneider 1989). In such functional models, different functions of scientific advice, for political actors, have been described—these functions are non-scientific, which means that science-based information is not used for its scientific insights but only for political reasons. Science-based information will only be used in a highly selective manner: If it does not conform to certain interests it will be neglected completely. In addition, scientific content ("epistemology") does not play any role for decision making, since scientific arguments are only assessed against the criterion of usefulness for political actors, and not of whether they are scientifically "state-of-the-art" or "true", etc.

These functional models are fruitful because they illustrate one important characteristic of politics: Most political decisions simply do not need scientific knowledge, and they are based only on political factors like interests, power, and path dependence. However, functional models cannot fully explain cases in which science-based advice influences or even forms political decisions by means of its epistemic features. The same models cannot fully explain cases in which "state-of-the art", science-based information becomes part of political decisions.

4.5 Co-production and Communication Models

Some models of scientific knowledge transfer are based on the idea that there is some kind of interaction between science and politics that leads to political decisions that reflect both scientific and political issues. In the co-production model by Hulme (2009), scientific and political arguments commonly influence policy making (Hulme 2009, p. 104). The co-production model represents the idea that achieving complex policy goals requires joint considerations of scientific and non-scientific arguments (Edelenbos et al. 2011; Hulme 2009, p. 104). Jasanoff notes that in the co-production model, scientific knowledge is "incorporated into practices of […] governance […]" (Jasanoff 2004, p. 3). Politics, in turn, also influences the actual use of scientific knowledge (Jasanoff 2004, p. 3). Co-production models emphasise that the utilization of scientific knowledge is not independent from the policy process. The terms "politicization of science" and "scientization of politics" (Maasen and Weingart 2005, p. 4) express how, contrary to a linear or systems theory explanation of scientific knowledge transfer, there are numerous "interactions between policymakers and experts" (Maasen and Weingart 2005, p. 7) that influence the use of scientific knowledge. Therefore, politicians often demand scientific expertise to realise their political goals, and political decisions thus influence the initiation of scientific research to supply this demand (Sarewitz and Pielke 2007). This model lies within the tradition of older approaches that emphasise the mutual cooperation between scientists and political actors, leading to political decisions that are based partly on science and partly on political considerations. It was German philosopher Jürgen Habermas who, in his normative "pragmatist" model of scientific policy advice, argued, as early as the end of the 1960s, that there cannot be a strict separation between the circles of scientific experts and those of political actors. For him, the relation between expertise and the political process is to be better understood as a mutually critical relationship of communication between science and politics. He argued that politicians and scientific experts come together in a mutually critical communication process within which scientific experts advice policymakers while the decision-makers, in turn, mandate the experts (Habermas 1969, p. 127). Experts are actively involved in the design, the derivation of alternative courses of action, to develop concrete measures and implementation strategies (Lompe 2006, p. 31). This communicative system involving scientific and political actors guarantees that new scientific findings and their consequences (like new technologies) can be permanently controlled by the value system of society and politics, whereas social interests are in turn controlled by the examination of technical and scientific possibilities and the strategic means of satisfying them (Habermas 1969, p. 127). The application of scientific knowledge in politics is controlled by the democratic decisions of society. Habermas' normative model of cooperation and communication between science and society was well received and advanced in the sociology of science (Weingart 2003, p. 94; Lompe 2006, p. 31) because his model of an iterative process of communication between experts and policy makers seems to be closer to reality than, for example, the aforementioned linear models, with their assumptions about the separation of knowledge and values, experts and politicians' roles (Weingart 2003, p. 94). It must be said that Habermas' assumptions about communication are very ideational, especially seen in the light of other theories mentioned above: Communication between science and politics is difficult due to the fundamentally different logic of either system. In addition, politics does not necessarily have an interest in communicating about scientific concepts.

Weingart has adopted Habermas' ideas in his recursive model of the science-policy interface. For Weingart, the science-policy interface should be understood as a non-linear, recursive communication process in which scientific experts communicate scientific issues and problems to political actors: Science helps to define policy problems and contributes to the agenda-setting process, often in collaboration with the media, and with politicians asking scientists for advice in finding solutions to these problems (Weingart 2003). Weingart helps to redefine the discursive link of Habermas under aspects of the role of expertise in governance. Discursive communication between political actors, experts and the media is important for an understanding of the role of expertise in politics. In his recursive model there is also space for the "scientization of politics" and "politicization of science", since the recursive process described takes place in a continuum between these two poles.

The different theoretical arguments presented by co-productionist and communicative models of scientific knowledge transfer highlight the important notion that it cannot be expected for political decisions to be fully based on scientific arguments (like in the linear or technocratic models) and that political decisions are formed by scientific arguments at least in part, and partly by political factors. These models appear to describe adequately the "real world" of science-policy interactions in modern societies: Experts do not frame political decisions autonomously and scientific expertise is used not only to serve political interests without any epistemic function. Rather, political decision making is the result of co-production between society, experts, and politicians. In addition, the communicative and recursive models show that, within political processes, political actors and scientists come together to exchange arguments: Especially in environmental policy, expert panels or commissions, round tables, advisory boards for political institutions, etc., represent such institutionalised forms of expert/politician communication. However, these models do not shed much light onto the micro-level processes within these science-policy interactions. It is said that there is communication or interaction and that such processes are shaped by political and scientific rationality, but what happens exactly at the interaction level remains somewhat unclear: How can communication and interaction function if scientists and politicians come from societal systems that differ fundamentally? Which activities lead to successful "co-production" and political decisions that reflect the scientific state of the art? It is difficult in these models to distinguish exactly between what is science and what is politics, since everything seems to be blurred in communication or co-production processes.

4.6 The RIU Model

So far, we have shown that every theoretical approach presents some important strengths, but also that none explains scientific knowledge transfer in a satisfactory way under the conditions of the political process. Linear/technocratic models can explain well the scientific influence on politics, in which science plays an epistemic role. But these models lack a deeper incorporation of politics and cannot explain why, in many cases, political actors do not utilize science at all or why political solutions deviate from the best scientific recipes. In contrast, functional models are good at explaining political aspects of scientific knowledge transfer, but they lack explanatory power in situations in which political solutions are based at least partly on science and in which science plays an epistemic role that goes beyond being just a scapegoat for politicians' interests. The merit of systems theory is that it shows the main differences between science and politics that lead to the conclusion that there is nearly no scientific knowledge transfer possible, due simply to the impossibility of communication between the two systems. The problem of such theories is that they cannot really explain cases in which scientific knowledge transfer can be observed. Co-productionist and communication-oriented theories are good for highlighting the mutual interactions between science and politics in modern society. However, they lack a deep analysis of the characteristics of these interactions and of how political and scientific factors are incorporated and communicated during these interactions despite the fundamental differences in logic between science and politics (see Table 4.1 for the main strengths of each approach).

Based on this brief review of some of the main theories of scientific knowledge transfer in political processes, we aimed to develop a scientific model for explaining scientific knowledge transfer that reflects the following main assumptions:

Table 4.1 Strengths of four main theoretical models of scientific knowledge transfer

Theoretical models	Strength in explaining…			
	Successful scientific knowledge transfer	Failed scientific knowledge transfer	Political aspects of scientific knowledge transfer	Epistemic function of science in scientific knowledge transfer
Linear/technocratic models	o	––	––	++
Functional models	o	++	++	––
Systems theory models	––	++	++	––
Co-production/communication models	o	o	o	o

(++ strong, –– weak, o some strength)

(1) The main precondition for scientific knowledge transfer is that it should be based on state-of-the-art science and that science should play an epistemic role for political and practical decisions.
(2) Political processes are based on political factors, like power and interests. These factors are the main influences for scientific knowledge transfer in the policy process.
(3) The science system and the political system function using a completely different type of logic. Scientific knowledge transfer happens despite the fact that political actors do not understand science and scientists do not understand politics.
(4) Communication between actors from science and politics is necessary but not sufficient for scientific knowledge transfer.
(5) A unique form of communication is necessary that reflects but does not change the unique rationalities of either system. This kind of communication does not change scientific rationality and does not try to change political power relations.

These assumptions reflect important findings from the theories presented above as well as findings from political science about the characteristics of political processes.

Our analytical RIU model defines the process of scientific knowledge transfer as the continuous process of research (R), integration (I), and utilization (U), with each following an individual logic (see chapters above).

Research is understood as the production of specific knowledge by using scientific methods and standards (Stevanov et al. 2013). The model reflects that high-level research must be regarded as an important precondition for each scientific knowledge transfer (Lentsch and Weingart 2011), because the quality of scientific expertise fundamentally determines its potential for effectively solving problems and serving interests. This is important to generate credibility and encourage utilization among practitioners (Pregernig and Böcher 2012, p. 204 f.). The potential for the use of scientific findings in environmental policy increases if these findings rely on "state-of-the-art" science. Therefore, like many other scholars, we argue that the quality of the scientific expertise produced is generally important for scientific knowledge transfer in policymaking (Lentsch and Weingart 2011).

Integration is defined as a bi-directional step in the process of interaction between practical utilization and research. During integration, research results that are relevant for political actors are selected using criteria based on practical demands. Meanwhile, practical demands for scientific solutions are translated into scientific research questions (Böcher and Krott 2010, p. 37). Integration is the exchange of information in response to the demands of those in practice and in evidence gathering; this process directs research toward practical problems with the aim of describing and successfully solving them (Böcher and Krott 2010, p. 37). Integration connects scientific evidence with the expectations and wishes of politicians without altering or changing the results of scientific research. The interplay between research and integration requires broad knowledge of the available scientific expertise and knowledge of political practice (Böcher and Krott

2010, p. 37). Integration connects research with the demands of political practice by selecting individual "bricks of knowledge." These bricks are valid because they reflect "state-of-the-art" science and are simultaneously politically relevant (Böcher and Krott 2010, p. 38). In this sense, these bricks represent the idea of "serviceable truth" as observed by Jasanoff (for a detailed discussion, see also: Guston 2001; Jasanoff 1990, p. 250). Serviceable truth passes tests of scientific evidence while also supporting political decision making (Jasanoff 1990, p. 250).

Utilization is the active use of scientific knowledge by actors in practice. Actors decide by themselves what kind of knowledge and how much of it they want to use, and they can also force other actors to use this scientific knowledge. In the RIU model, multiple power relations exist that support the utilization of scientific knowledge in political practice. These multiple power relations support multiple science-based solutions.

4.6.1 Multiple Power Relations Supporting Multiple Science-Based Solutions

We argue that, in society, power relations exist in multiple relations between different stakeholders and in multiple societal networks. We follow a state theory with the idea of the existence of polycentric actor configurations and not the classic idea of a unitary powerful hierarchical state (Rieger and Schultze 2002, p. 493). Reflecting classic models of power (Weber 1972), we argue that in society there exist a nearly unlimited number of micro power relations, in which one person is able to influence the actions of one or more other persons, to force those others to act in the interest of the person exerting the influence (French 1956, p. 182). Our understanding of power is based on the actor-centred power approach developed by Krott et al. (2014, p. 37): "actor-centred power is a social relationship in which actor A alters the behaviour of actor B without recognising B's will", a definition that is based on classic approaches by Weber (2000) and Dahl (1957). "The model considers a social relationship between actor A and actor B. In most cases both actors try to alter each other's behaviour, and one actor resists the other's efforts to a degree." (Krott et al. 2014, p. 37). In the actor-centred power approach the actor who alters the behaviour of another actor is called "potentate" and the other actor "subordinate" (Krott et al. 2014, p. 37). "Every actor plays, in different social relations, sometimes the role of potentate and sometimes the role of subordinate, depending on the specific issue" (Krott et al. 2014, p. 37). In the relationship of power can be such actors as individuals, groups, roles, offices, governments, nation-states etc. (Dahl 1957, p. 203; Krott et al. 2014).

In politics and society many of those power relations exist, for example, in enterprises where the owners control the management by using the potential of power. These multiple power relations offer to the more powerful actor the option of using power to introduce scientific rationality even against the will of other actors. In this way, we see power as a very important driving force for the transfer of scientific information. Our argument is that, every time one actor has power over another actor he (the potentate) is able to force the latter (the subordinate) to use scientific information. This model of scientific knowledge transfer can be named “science plus power”. Science is used here as an information source, and it has an epistemic function. The information is acquired and selected by actors who add power in order to force others to apply scientific rationality. Science does not support politics, but politics support science. To elaborate this main argument of our RIU model further, in the following section we discuss four types of politically-driven scientific information transfer based on different power relations of “science plus power”. Each of these power relations are based on a specific actor who serves as an important “ally for science”.

4.6.2 “Science Plus Power”: Four Powerful Allies for Science in the RIU Model

Our basic argument is that the transfer of scientific information into political practice is caused by power that is added to the selection of scientific arguments (“science-plus-power model”). Scientific arguments need an actor as an ally who has the necessary power at his disposal that creates application of science-based solutions by other actors and leads to the transformation of science into action. We distinguish four different types of allies, who each make use of science differently. The four types of actors are “internal allies”, “external allies”, “learning allies”, and “wise allies”. These types of allies can be distinguished by the form of their internal or external participation in the research process. The allies differ in the way they evaluate and/or change the scientific rationality in transferred information. Table 4.2 provides an overview of the four types of allies, with their main characteristics.

Table 4.2 The four types of allies in the RIU model

Allies in the RIU model	Definition	Scientific rationality of transferred information	Internal/external to the research process
Internal ally	Takes part in the research process and supports dissemination of selected results with power	Not evaluated not changed	Internal
External ally	Does not take part in the research process, but pressures other actors to select and use scientific results	Not evaluated not changed	External
Learning ally	Selects scientific results to adapt his own interests to them	Not evaluated not changed	Internal or external
Wise ally	Selects and changes scientific results based on a wide range of knowledge	Evaluated and changed	Internal or external

4.6.3 Internal Allies

One important mechanism for transforming scientific information into practical application is the political production of acceptance through "internal allies" of science. In the model of "internal allies" political actors participate internally in the research process and support the dissemination of selected results by means of power. They utilize scientific discourses directly by judging scientific arguments against the background of underlying political criteria. Political actors assess scientific arguments with regard to their individual interests and political goals. They apply scientific solutions because these promise them advantages against other actors and serve to identify and realise their interests. Internal allies employ all available resources so that other actors also utilize the scientific solution. Political actors serve as internal allies for selected scientific information, but the political actors do not evaluate or change the scientific rationality.

The case study "Hand in hand—Life 2014 in the Pinzgau region" (Sect. 3.2) gives a good example for the relevance of internal allies. Here, local mayors and interns were part of the research project and were able to influence other actors to the use science-based solutions for regional development.

4.6.4 External Allies

In the "external ally" type, political actors are not directly involved in the research process and the dissemination of its results. The external ally exerts pressure on other actors with the aim of leading them to cooperate with scientific advisers and to coerce them to utilize scientific knowledge. An external ally of science coerces other actors to solve their problems autonomously via the help of scientific advice and by using scientific information. This mechanism can be useful for the stronger actor since other actors might solve problems in a better way if they use scientific knowledge. So the stronger actor gains advantages due to the fact that he does not have to invest his own resources to find better political solutions. The powerful actor has a strong interest in increasing the rationality of the actions of the controlled actors in order to avoid excessive investment of his own resources. Political actors serve as external allies for selected scientific information but the political actor does not evaluate the scientific information or change the scientific rationality.

The case study "Establishing a legal basis: Implementing the EU Water Framework Directive in Austria" (Sect. 3.3) gives a good example for the relevance of external allies. Here, the European Union were able to force other actors to develop and to use science-based concepts that are necessary to implement the EU Water Framework Directive.

4.6.5 Learning Allies

Another type of ally for scientific knowledge transfer is the "learning ally". Learning allies are actors who learn something new from science concerning their interests. As consequence, they reinterpret and/or adjust their interests in light of new scientific insights. This model reflects the idea of modern public policy analysis that says that political actors are able to learn and do not always follow strictly rationally their exogenously constituted short-term interests. Learning means that new experiences had by the actors lead to consequences for current or future acting (Bennett and Howlett 1992). An increase of available information due to scientific advice can reach actors in the political system and lead to learning processes (Majone 1989). Normally, political actors assess scientific insights against the background of their interests and in search of advantages. With this type of ally, scientific knowledge puts a new light on these interests in a way such that actors have to re-interpret their interests and rethink their position even if they are driven purely by egoistic motivations.

Specifically for the learning approach, we underscore the limits set by the powerful actor. He is able to learn but this learning is always egoistic. He will not act altruistically in order to support political goals. But he will adapt his interpretation of his own interests, integrating scientific arguments, and will enforce the results against less powerful actors. Learning is very demanding for the actor but, due to the limited information, not impossible, even if we assume the dominance of orientation toward profits in the short term. In this sense, the learning of a powerful actor is limited but has high relevance for less powerful actors, whereas learning by a less powerful actor can help said actor to identify his best interest within the framework set by more powerful actors. Learning allies are external or internal to the research process and do not evaluate the scientific information or change the scientific rationality.

4.6.6 Wise Allies

From the point of view of the political actor, scientific knowledge is just one kind of relevant information source that has to be assessed in order to find the necessary problem-solving information. Therefore, practical decisions are also based on information produced by other knowledge sources like tradition, stable belief systems (Sabatier and Jenkins-Smith 1993), practical experiences, or lay or tacit knowledge (Bäckstrand 2003). The political actor is confronted with the difficult challenge of evaluating all the different knowledge sources and of assessing all of these possible parts of the information "puzzle". For this judgment, science is not very helpful since science cannot assess the quality of other kinds of knowledge.

Science itself only has a potential for scientific argumentation, without being able to assess other forms of thinking. Wise allies evaluate the significance of scientific knowledge within the entirety of issue-related information available.

These actors are wise in the sense that they know how to incorporate scientific findings correctly into the decisions they have to take. The wise ally's decision is also lifeworld-oriented. The rules of lifeworld go far beyond scientific rules so that science only plays a limited role in producing the overall information. However, it cannot in any way be the aim of science to produce this knowledge as a whole since in that case science would lose its specific characteristic—the search for "truth" guided by specific rules, against the background of specific criteria and evaluation by scientific peers.

The mechanism of the "wise ally" reflects the idea of the importance of new forms of producing knowledge (Gibbons et al. 1994) that are characterised by a loss of importance of science as the single source of knowledge production and by an expansion of different knowledge producers and users in society other than science (Gibbons et al. 1994, p. 167). It is also reflected in the idea of an agora (Nowotny et al. 2001, p. 201) in which this image from ancient Greece is used to describe a new public space where "science and society, the market and politics co-mingle" (Nowotny et al. 2001, p. 203), which in our interpretation means that different sources of information are always mixed to find political decisions and that science is no longer merely an external force that can be brought into decision-making processes by demand. Rather, science is an omnipresent source of information in the agora (Nowotny et al. 2001, p. 206), whose relevance has to be assessed permanently by political actors, in contrast to other knowledge sources. Science here is not the one and only information source but an important source of knowledge that complements, and is complemented by, other knowledge sources.

The wise ally is an actor who selects science twice: First, he uses (lifeworld-based) arguments other than scientific arguments to supplement and change scientific findings, and second, he applies his own interests and power to achieve an application of his preferred arguments, which are then constituted partly by science and partly by the lifeworld. In our model, "science plus power", the wise ally is the most problematic actor type because he changes scientific arguments and selects information based on interests and power. In this regard, the appropriation of science by the wise ally is based on scientific rationality to a lesser degree than is the case with the other types of ally. This is because the wise ally not only selects the research results without verifying their scientific quality, but also even changes the validity of these results in that he mixes them with lifeworld arguments. Therefore the wise ally is the least scientifically based actor of all types within the "science plus power" model, because he selects and changes scientific results based on a wide range of knowledge.

4.6.7 Power for Scientific Knowledge Transfer Against Democratic Rules?

We suggest in our RIU model that scientific knowledge needs powerful allies who support certain scientific solutions in their realm. They do this without changing scientific rationality because the scientific solution serves their interests. They can do this without having to be able to understand the scientific basis of a certain solution. Scientific knowledge transfer takes place despite the fact that the types of logic of the two systems, science and politics, are different. It can take place even if communication or translation from one system to another is nearly impossible.

The main assumption of RIU, that power drives scientific knowledge transfer, seems to be undemocratic and biased. But we think that due to the multiple power relations in society and due to changing political governments and political moods, our RIU model does reflect the reality of modern democratic political systems to a great extent: In many cases, formerly weak actors can become powerful, so that a given science-based solution they may favour and for which they serve as an ally can gain power as well, and can be realised. Therefore, the RIU model reflects non-permanent and multiple power relations in society that are always changing.

Let us end again with the example of the change that took place in German nuclear energy policy after the Fukushima catastrophe in 2011: The German government re-assessed the risks of nuclear power and decided to decommission German nuclear power plants and start a strategy to promote renewable energy use in Germany. Under these changed conditions, formerly weak allies of scientific solutions for renewable energy systems, like environmental NGOs, were able to gain strength as important allies of those new science-based solutions. Scientific concepts for renewable energy use were already well known years ago; what was yet needed were political changes in the power relations for the utilization of alternative science-based solutions that existed long before. Utilization of science is here by no means a question solely dependent on scientific rationality. The example shows that, due to the fact that even weak interests can become stronger due to changing political conditions and that many different power-relations exist at the same time, our RIU model does reflect the reality of modern democratic society not necessarily leading to a biased (ab)use of science. Science will "enlighten" neither politics nor society by speaking "truth to power", but if science speaks "truth within power" and links this to actors professionally, as the RIU model describes, then knowledge transfer can be successful.

References

Ascher, W., Steelman, T., & Healy, R. (2010). Knowledge and environmental policy: re-imagining the boundaries of science and politics. Cambridge, MA: MIT Press.

Bäckstrand, K. (2003). Civic science for sustainability: Reframing the role of experts, policy-makers and citizens in environmental governance. *Global Environmental Politics, 3* (4), 24–41.

Beck, S. (2011). Moving beyond the linear model of expertise? IPCC and the test of adaptation. *Regional Environmental Change, 11*, 297–306.

Beck, U. (1992). *Risk society: Towards a new modernity* (Vol. 17). London: Sage.

Bennett, C. J., & Howlett, M. (1992). The lessons of learning: Reconciling theories of policy learning and policy change. *Policy Sciences, 25*(3), 275–294.

Böcher, M., & Krott, M. (2010). Umsetzung des Konzepts einer modernen Ressortforschung im Geschäftsbereich des BMU (Vol. 39/2010, Umweltbundesamt-Texte). Dessau-Roßlau: Umweltbundesamt.
Böcher, M., & Töller, A. E. (2012). *Umweltpolitik in Deutschland: eine politikfeldanalytische Einführung* (Vol. 50). Wiesbaden: VS Springer-Verlag.
Böcher, M., & Töller, A. E. (2015). Inherent dynamics and chance as drivers in environmental policy? An approach to explaining environmental policy decisions. Paper presented at the International Conference on Public Policy, Milan,
Bocking, S. (2004). *Nature's experts: Science, politics, and the environment*. New Brunswick, New Jersey: Rutgers University Press.
Boehmer-Christiansen, S. (1995). Reflections on scientific advice and EC transboundary pollution policy. *Science and Public Policy, 22*(3), 195–203.
Bradshaw, G. A., & Borchers, J. G. (2000). Uncertainty as information: Narrowing the science-policy gap. *Conservation Ecology, 4*(1), 7.
Braun, D., & Benninghoff, M. (2003). Policy learning in Swiss research policy—the case of the National Centres of Competence in Research. *Research Policy, 32*(10), 1849–1863.
Dahl, R. A. (1957). The concept of Power. *Behavioral Science, 2*(3), 201–215.
Dunn, W. N. (2015). *Public policy analysis* (5 ed.). London, New York: Routledge.
Durant, D. (2015). The undead linear model of expertise. In M. Heazle & J. Kane (Eds.), *Policy legitimacy, science and political authority: Knowledge and action in liberal democracies* (pp. 17–38). Abingdon, New York: Routledge.
Edelenbos, J., van Buuren, A., & van Schie, N. (2011). Co-producing knowledge: Joint knowledge production between experts, bureaucrats and stakeholders in Dutch water management projects. *Environmental Science & Policy, 14*(6), 675–684.
French, J. R. P. J. (1956). A formal theory of social power. *Psychological Review, 63*(3), 181–194.
Gibbons, M., Limoges, C., Nowotny, H., Schwartzman, S., Scott, P., & Trow, M. (1994). The new production of knowledge: the dynamics of science and research in contemporary societies. London [et. al.]: Sage Publications.
Grundmann, R. (2009). The role of expertise in governance processes. *Forest Policy and Economics, 11*, 498–503.
Gupta, A., Andresen, S., Siebenhüner, B., & Biermann, F. (2013). Science networks. In F. Biermann & P. Pattberg (Eds.), *Global environmental governance reconsidered* (pp. 69–93). Cambridge, MA: MIT Press.
Guston, D. H. (2001). Toward a "best practice" of constructing "serviceable truths". In M. Hisschemöller, R. Hoppe, W. N. Dunn, J. R. Ravetz (Eds.), *Knowledge, power, and participation in environmental policy analysis. Policy Studies Review Annual No. 12* (pp. 97–118). New Brunswick/London: Transaction Publishers.
Haas, P. M. (2005). Science and international environmental governance. In P. Dauvergne (Ed.), *Handbook of global environmental politics* (pp. 383–401). Cheltenham: Edward Elgar.
Habermas, J. (1969). Technik und Wissenschaft als "Ideologie" (1. - 15. Tsd. ed., Edition Suhrkamp; 287). Frankfurt am Main: Suhrkamp.
Hulme, M. (2009). *Why we disagree about climate change: Understanding controversy, inaction and opportunity*. Cambridge: Cambridge University Press.
Jasanoff, S. (1990). *The fifth branch: Science advisers as policymakers*. Cambridge, Mass: Harvard Univ. Press.
Jasanoff, S. (2004). The idiom of co-production. In S. Jasanoff (Ed.), *States of knowledge: The co-production of science and social order* (pp. 1–12). London: Routledge.
Jasanoff, S., & Wynne, B. (1998). Science and decisionmaking. In S. Rayner & E. L. Malone (Eds.), *Human choice and climate change* (pp. 1–87). Columbus, Ohio: Battelle Press.
Kingdon, J. W. (2003). Agendas, alternatives, and public policies (2nd ed., Longman classics in political science). New York [u.a.]: Longman.
Kirchgässner, G., & Schneider, F. (2003). On the political economy of environmental policy. *Public Choice, 115*(3–4), 369–396.

Krott, M. (2012). Value and risks of the use of analytical theory in science for forest policy. *Forest Policy And Economics, 16*, 35–42.

Krott, M., Bader, A., Schusser, C., Devkota, R., Maryudi, A., Giessen, L., et al. (2014). Actor-centred power: The driving force in decentralised community based forest governance. *Forest Policy and Economics, 49*, 34–42.

Lehmann, L., & Rieder, S. (2003). Von Mägden und Knechten in Wissenschaft und Politik. tec21 (26), 6–8.

Lentsch, J., & Weingart, P. (Eds.). (2011). *The politics of scientific advice. Institutional design for quality assurance*. Cambridge: Cambridge University Press.

Lompe, K. (2006). Traditionelle Modelle der Politikberatung. In S. Falk, D. Rehfeld, A. Römmele, & M. Thunert (Eds.), Handbuch Politikberatung (pp. 25–34). Wiesbaden: VS Verlag für Sozialwissenschaften.

Luhmann, N. (1968). Selbststeuerung der Wissenschaft. Jahrbuch für Sozialwissenschaft (H. 2), 147–170.

Luhmann, N. (1981). Theoretische Orientierung der Politik. In Soziologische Aufklärung (Ed.), *3* (pp. 287–292). Opladen: Westdeutscher Verlag.

Maasen, S., & Weingart, P. (2005). What's new in scientific advice to politics? In S. Maasen, & P. Weingart (Eds.), Democratization of expertise? Exploring novel forms of scientific advice in political decision making (Vol. 24, pp. 1–19, Sociology of the Sciences). Dordrecht: Springer.

Majone, G. (1989). *Evidence, argument and persuasion in the policy process*. New Haven, London: Yale University Press.

Mayntz, R. (1983). Lessons learned: Problems in the acceptance of TA by political decision-makers. In Umweltbundesamt (Ed.), Technologien auf dem Prüfstand: Die Rolle der Technologiefolgenabschätzung im Entscheidungsprozeß; die Beiträge zum Internationalen Symposium vom 19. bis 21. 10. 1982 in Bonn (pp. 333–345). Köln: Heymanns.

Miller, N. (2008). *Environmental politics: Stakeholders, interests, and policymaking*. New York, London: Routledge.

Mueller, D. C. (2004). Public choice: An introduction. In C. K. Rowley, & F. Schneider (Eds.), The encyclopedia of public choice (Vol. 1, pp. 32–48). Dordrecht [u.a.]: Kluwer Academic Publishers.

Nowotny, H., Scott, P., & Gibbons, M. (Eds.). (2001). Re-thinking science. Knowledge and the public in an age of uncertainty. Cambridge [u.a.]: Polity Press [u.a.].

Nutley, S. M., Davies, H. T., & Smith, P. C. (2000). What works? Evidence-based policy and practice in public services. MIT Press.

Parsons, W. (1995). *Public policy*. Cheltenham, Northampton: Edward Elgar.

Pregernig, M. (2014). Framings of science-policy interactions and their discursive and institutional effects: Examples from conservation and environmental policy. *Biodiversity and Conservation, 23*(14), 3615–3639.

Pregernig, M., & Böcher, M. (2012). Normative and analytical perspectives on the role of science and expertise in environmental governance. In K. Hogl, E. Kvarda, R. Nordbeck, & M. Pregernig (Eds.), *Environmental governance: The challenge of legitimacy and effectiveness* (pp. 199–219). Cheltenham: Edward Elgar.

Radaelli, C. M. (1995). The role of knowledge in the policy process. *Journal of European public policy, 2*(2), 159–183.

Rieger, G., & Schultze, R.-O. (2002). Machttheoretische Ansätze. In D. Nohlen, & R.-O. Schultze (Eds.), Lexikon der Politikwissenschaft. Theorien, Methoden, Begriffe, (pp. 488–495). München: Beck.

Sabatier, P. A., & Jenkins-Smith, H. C. (1993). *Policy change and learning: An advocacy coalition approach (Theoretical lenses on public policy)*. Boulder, Colorado.: Westview Press.

Sarewitz, D. (2004). How science makes environmental controversies worse. *Environmental Science & Policy, 7*(5), 385–403.

Sarewitz, D., & Pielke, R. A, Jr. (2007). The neglected heart of science policy: Reconciling supply of and demand for science. *Environmental Science & Policy, 10*, 5–16.

Schelsky, H. (1965). Der Mensch in der wissenschaftlichen Zivilisation. In H. Schelsky (Ed.), *Auf der Suche nach Wirklichkeit* (pp. 439–480). Düsseldorf: Bertelsmann Universitätsverlag.

Schneider, W. L. (1989). Kooperation als strategischer Prozeß. Administrative Auftragsforschung im Spannungsfeld zwischen professionellem Interesse und politischer Instrumentalisierung. In U. B. Beck, Wolfgang (Ed.), Weder Sozialtechnologie noch Aufklärung? Analysen zur Verwendung sozialwissenschaftlichen Wissens (Vol. 302–331). Frankfurt/M.: Suhrkamp.

Stevanov, M., Böcher, M., Krott, M., Krajter, S., Vuletic, D., & Orlovic, S. (2013). The Research, Integration and Utilization (RIU) model as analytical framework for the professionalization of departmental research organizations: Case studies of publicly funded forest research institutes in Serbia and Croatia. *Forest Policy And Economics, 37*, 20–28. doi:10.1016/j.forpol.2013.03.006.

Warren, M. E. (1999). What is political? *Journal of Theoretical Politics, 11*(2), 207–231. doi:10.1177/0951692899011002004.

Weber, M. (1972). Wirtschaft und Gesellschaft (5 ed.). Tübingen: Mohr.

Weber, M. (2000). *Basic concepts in sociology*. New York: Citadel Press.

Weingart, P. (2003). *Wissenschaftssoziologie (Einsichten: Themen der Soziologie)*. Bielefeld: Transcript-Verl.

Willems, H., Wolf, M., & Eckert, R. (1993). Soziale Unruhen und Politikberatung. Funktion, Arbeitsweise, Ergebnisse und Auswirkungen von Untersuchungskommissionen in den USA, Großbritannien und der Bundesrepublik. Opladen: Westdeutscher Verlag.

Chapter 5
Outlook—Further Potential Applications of the RIU Model

5.1 From Analyzing to Improving Scientific Knowledge Transfer

The RIU Model is an analytical tool for the identification of the critical factors that shape the transfer of scientific knowledge to policy making. Therefore, in addition to aiding the analysis of scientific knowledge transfer cases, the RIU model can help improve knowledge transfer and its evaluation.

The normative goal resulting from a consequent use of RIU and its criteria is state-of-the-art research that provides exactly the knowledge political actors need in order to make the best decisions in the course of their public missions. For meeting this goal we can derive from RIU the ideals of a connection between high quality scientific research, pluralistic integration and democratic political utilization as providing a fruitful normative frame.

5.1.1 Independent High Quality Scientific Research

Scientific research follows its own logic. The tasks of science are grounded in empirical evidence so as to describe and explain the world based on theories. This vast production of sound and innovative scientific knowledge can work properly only if scientific logic dominates and scientists do not have to face intervention from political practice. Scientists must be free to ask all questions and communicate all results within the scientific discourse, independently from whether the results serve the interests of any actors in politics, the economy or society.

Securing "good scientific practice" and the state-of-the-art quality of scientific work are the main pillars to guarantee sound research procedures. This means that scientists make statements both about results upon which they all agree as well as about those where there is disagreement. The latter is normal because scientific

M. Böcher and M. Krott, *Science Makes the World Go Round*,
DOI 10.1007/978-3-319-34079-1_5

research can produce contradictory results. In such cases, it makes no sense, scientifically, to force scientists to vote for one common solution. The fact is that those questions that have led to contradictory results have to remain open, because a decision as to what the scientific answers to them should be cannot be made at a particular moment. This is a standard result of good scientific practice in these cases. Scientists must have the freedom to be specific, and to admit what they do not know, even if the policy issue in question is a matter of survival. Even scientists' best guess is of no scientific value at all, and relying on such guesses endangers the credibility of scientists if these guesses turn out to be wrong later.

Furthermore, scientists must be free to make statements the veracity of which can only be demonstrated within a specific degree of probability, very often the case in science. Such statements have high scientific value, but for political and practical actors they mean that, although the scientist knows something about the problem, there is still a risk that the scientific statement may be wrong. A scientific statement based on such probabilities can never be transformed into consensual, clear, scientific advice. The decision as to how seriously the risk should be regarded remains outside the scientific system and has to be made by the political and practical actors responsible for practical and political decisions.

In summary, within excellent scientific research there has to be a distinction between (1) proven results, (2) results having a certain probability of being right and (3) the unknown. If scientists have the freedom to produce any of these kinds of results without interference or political pressure, they will be able to maintain high quality in their research and to remain innovative, gaining as a consequence high public credibility. An important prerequisite for this is that scientists need to be protected from any direct intervention by actors in politics, the economy or society, as the RIU model shows.

5.1.2 Professional, Transparent and Pluralistic Integration

The RIU Model shows that the transfer of scientific knowledge depends on the selection of specific results and on the support the results get from political and practical actors. For scientific results to be transformed into policy-making, scientists have to receive maximum political support for their findings. In this process, different actors acquire different amounts of scientific knowledge. Making the winners and losers of a science-based solution visible is a requirement for professional integration in order to support a balanced pluralistic system.

Scientists must realise that, as far as integration is concerned, producing innovative scientific solutions, by itself, does not change practice at all. A necessary follow-up step is to win allies who are willing and able to implement the solutions in practice. Professional integration means selecting the research questions and the results correctly in order to gain maximum support from political actors. This works only if scientists have specific public or private actors in mind, because society and the economy as a whole do not take action.

A consequence of the differentiation of actors is that some actors get much from science and some get nothing, because they cannot communicate with science, they are too weak and they cannot find allies for their interests. Integration intervenes in the power balance of society by supplying specific actors from the economy, politics and society who have scientific knowledge. According to the RIU model, independent scientific research as such does not intervene. Intervention results from integration, as a separate element within the process of transformation, or scientific knowledge transfer, that achieves change. To fulfil the ethical standards of pluralistic societies, integration must reveal value-laden and interest-driven intervention, making the public goals and allies visible.

Specific integration will always be biased toward specific public goals, actors and their interests. Professional integration identifies many potential actors and has the means to communicate the scientific results to them. These multiple contacts open up decision space for selected allies. Nevertheless, the selection of allies itself cannot be decided scientifically, and will be biased with regard to active allies. The existence of such a bias is accepted within a pluralistic system because the diversity of interests is at the very heart of pluralism. This diversity of interests and different allies guarantee that checks and balances emerge that weaken the biases. Additionally, democratic elections and their results lead—at least theoretically—to changing phases of political support for certain allies.

The great diversity of actors and their different power sources open many spaces for scientific knowledge to be utilized. Whenever a specific actor has more power than another actor he can use his power sources as support for a specific scientific solution and is able to force the other actor to follow this solution. The result is that scientific research results are transformed into practical solutions. By recognising that power can support the transfer of scientific knowledge, the RIU model draws attention to the many power differences in society and in politics. Each difference opens an opportunity for knowledge transfer. No single strong government or company has a monopoly of power to force knowledge transfer, but nearly all actors contribute to knowledge transfer, insofar as they are able to dominate other, specific actors.

Therefore our odd message that power decides the transfer of knowledge and not scientific arguments comprises additionally the good news that due to the omnipresence of different power relations in society and politics also weaker actors can get chances to implement scientific knowledge if they are able to find stronger actors that support this knowledge by using their power sources.

Nevertheless, exposing the bias toward specific interests helps to maintain a fair, pluralistic and competitive discourse. The media also have to make their connections to different interests transparent. Finally, the pluralistic discourse needs there to be tolerance for diverse interests in society and politics.

5.1.3 Binding, Responsible and Democratic Political Decisions

The final selections of problems and solutions are decisions that are the duty of the political system. Whereas professional integration actively intervenes in the public discourse, the final selection of problems and solutions is a genuine political task that cannot be carried out by science.

The function of the political system is to organise collective action in the context of differing short-term interests and the power relationships that exist in modern society (Warren 1999). Even in spite of different interests, politics has to produce binding decisions about which problems should be solved in which manner. Here, "binding" means that society and the economy have to accept political decisions and follow the prescribed solutions, even if specific actors and groups prefer the alternatives. The political system has the power to implement the binding solutions. Democratically elected political actors have the mandate to act in this binding way, but, in addition, voters hold them responsible for decision-making and implementation. If the voters are unhappy with the consequences of these decisions they are free to choose other politicians in the next elections. From the perspective of democracy, this is one important reason why it is politicians who should make decisions on science-based solutions, and not scientists: Scientists do not have the mandate to make decisions concerning the problems of society, they are not democratically legitimised by voters and they cannot be held responsible if a decision fails. This is one important reason why, for example, even in the face of global problems, like climate change, no scientific authority can make these decisions: Such decisions can only be produced by the political system, even if they deviate substantially from scientific rationality.

Therefore, the RIU model stresses that it is important that the decisions made by those in positions of responsibility should follow political procedures based on law. Political actors are always active in deciding on political options, where no decision can be made based on scientific facts alone. Scientific expertise that generates factual information therefore never suffices for decision-making and cannot "force" politicians to act as they do. Politicians must not delegate these decisions to science, and they always have to retain responsibility, since they are the only actors that can be held accountable directly and in a democratic manner.

Making decisions and assigning responsibility for them is necessary for a democratic system. Democracy does not imply a general political consensus, but rather, that politicians are given a democratic mandate to make decisions about conflicting issues. Democratic procedures hold the politicians responsible and can lead to an end of the mandate as a consequence of the decisions. Democratic procedures require transparency and the participation of society, supported by successful integration of scientific knowledge, but the final decisions remain a duty of the political system.

To summarize, the RIU model provides a basis from which to strengthen the transfer of scientific knowledge in a manner that is guided by the ideals of high-quality independent scientific research, pluralistic integration, and democratic political decision-making.

5.1.4 Some Brief Cases Seen Through the Lens of the RIU Model

How RIU works to inform practice is discussed in this chapter, against the background of some selected current cases and discussions about scientific knowledge transfer. The national (German) and international cases presented briefly here do not offer an elaborate programme for reforming a specific organisation or procedures. Rather, they may suggest what improvements based on the RIU model would look like. To push through the reforms we sketch below, a full "RIU process" would be necessary. Such a process would involve initiating professional integration to identify actors who are willing and able to support reforms, carrying out independent, high-quality research on possible solutions and making democratic decisions on the implementation of binding reforms.

5.2 Better Division of Tasks for a Scientifically Sound, Pluralistic and Democratic Intergovernmental Panel on Climate Change (IPCC)

In 1988 the World Meteorological Organization (WMO) established, in co-operation with the United Nations Environment Programme (UNEP), the Intergovernmental Panel on Climate Change (IPCC), which would serve as an international advisory body (Beck 2010, p. 15). The mission of the IPCC is to collect scientific research on climate change and to assess its state objectively, openly and transparently, including its observed and projected impacts and potential policy responses (regarding climate adaptation and climate mitigation options) (Beck 2010, p. 15). The IPCC delivered its first report in 1990 and four more reports since then, the final draft of the latest in 2014.

The major achievements of the IPCC are that its work helped to document continually the most current climate research. The latest report consists of more than 5000 pages, which is evidence of an enormous collective scientific effort (Nature Geosci 2013). The work of the IPCC was an important contribution to a better understanding of human-climate interactions. Its latest report "brings yet another confidence that human actions are altering the Earth's climate" (Nature Geosci 2013). Nevertheless, there has been criticism as well. Among the points for criticism is the fact that, despite its major achievements since its foundation, the

IPCC's reports have had only a limited impact on international climate policy—a global negotiation for substantial climate mitigation is still not in sight. Another point is the amount of work researchers have to invest to produce its massive reports. Some researchers argue that it would be better to work on smaller, issue-oriented assessments (Stocker 2013) than to write extensive reports inefficiently, of which only the synthesis report will be read by an audience beyond climate scientists. Furthermore, it is argued that the work of the IPCC could mix research and politics too much—especially in the production of the synthesis report, which government delegates have to approve before it can be published. Additionally, non-conformance with criteria of good scientific practice has also drawn criticism, because allegedly some of the scientific sources used by the IPCC in the past lack proper peer reviews. A further point is that the IPCC operates with probability only and experts claim that there is no further progress possible that goes beyond the current statement that the probability that global warming is human-induced is 95 % (Nature Geosci 2013). It does not seem to be possible to increase the probability associated with this consensus any further. In 2001 the probability was reported as 66 %; in 2007 it was 90 % (Nature Geosci 2013).

The criticism described here, originating in both politics and science, has caused some frustration among scientists. They claim that not all of the important scientific information available was included in the final "Summary for Policy-makers" that is part of the synthesis report. The prominent German researcher from the Potsdam Institute for Climate Impact Research (PIK), Ottmar Edenhofer, regretted that important ex-post emissions analyses were excluded due to political actors' fears that the national history of emissions would become a burden for specific countries in the political discourse in the future (Edenhofer and Minx 2014). Critically, Edenhofer and Minx claim that "a reasonable summary of ex-post policy analysis that considers different evaluation criteria is advantageous for negotiations" (Edenhofer and Minx 2014, p. 38). Despite that "factual and normative statements can usually be analytically distinguished", "the IPCC failed to agree on a reasonable way to analytically distinguish scientific analysis from potential political and legal interpretations" (Edenhofer and Minx 2014, p. 38). The authors warn that these failed practices will undermine the credibility of science in the future. Along these lines, a newer critique, raised by German political scientist Oliver Geden, drew the attention of climate scientists and the media in 2015 (Geden 2015): Geden argued that climate scientists should not become pragmatic and converge on political aspects in their scientific work. Even if there is no global political consensus in sight for reducing CO_2-emissions substantially, "scientific advisers need to resist pressure to fit the facts to the failure", but they have to "maintain integrity" (Geden 2015, p. 27). Otherwise, if they become part of the political system, they lose their scientific credibility. Geden's argument is in agreement with our RIU model: Even if scientists are unhappy with political decisions, they should not become political themselves—they should still deliver unaltered scientific facts, even if these are not accepted by politics. The responsibility for not acting has to be kept in the political system.

Recapitulating, in the past the IPCC served as a commendable critical voice within the climate change debate, but today scientists warn of a loss of the IPCC's credibility, and many political actors ignore the IPCC's recommendations in designing climate policy measures. The recent working practice and the design of the IPCC reveal failures in the transfer of scientific knowledge on climate change to the political process. When one looks at the IPCC through the lens of the RIU model, it indicates that there is potential for improvement in the IPCC's scientific knowledge transfer. According to the RIU model, the IPCC working process and its reports should discriminate between three tasks: high quality independent science, pluralistic integration and democratic decision. Three different reports accompanied by specific communication strategies could make the new approach visible to the public.

Report 1 would be on "Climate change science—the state of art", and it would present an overview of the most current findings in climate change science. It is important that the report should discriminate between results about which a scientific consensus exists and results that are contradictory. For example, German climate scientist Hans von Storch points out that there is a consensus about the warming of the climate and an increase in emissions, but that he finds that there are contradictions about the causative processes (von Storch 2014). Furthermore, the report should make the result probabilities known, especially in regard to the models for the future. It is also important to describe the most important questions to which science has no answers yet. Only the scientists involved in the IPCC's work are in charge of this report, and in doing this work they have to follow good scientific practice strictly and remain completely independent from any influence by political actors. Conversely, it follows that political actors have to accept any report (and any scientific results) and that they are not responsible for the scientific report.

Report 2 should be on "Options for climate change policy". This report should be selective with regard to preferred options and should not be built on an overall agreement. It should be produced through a pluralistic integration process. Scientists would look for active support from political actors. Within the IPCC, specific scientists would build alliances with specific political actors and design preferred options, making use of the selected scientific information. In addition, scientists would have to guarantee that all factual arguments that are selected are supported by scientific evidence and that all relevant probabilities are provided. Here, pluralistic responsibility is important. Different alliances of scientists and political actors would be responsible for each option, and not the IPCC as a whole. However, the IPCC would be responsible for divulging the links between the options and the specific ally coalitions. The report would provide decision makers with a scientifically sound and politically relevant road map of different options in climate change policy. Keeping the alternatives visible would best reflect the reality of pluralistic societies and political systems at international and national levels.

Report 3 would be on "Political recommendations" and would reflect joint decisions by all political actors involved in the IPCC, something that falls under their responsibility. Here, political decisions would be made about the assessment of specific alternatives as the best political options. The IPCC would organise a decision process and a vote based on its result. This is genuinely political and cannot be prescribed scientifically. For the political impact, it is important that the IPCC should produce joint political recommendations, which would be formulated by the IPCC as a whole. This means that it would be legitimate for a national political body to make a different political decision later, even if it participates in the IPCC. The joint recommendation is an important voice in the political discourse but it cannot overrule the democratic national decisions of each individual country. The political report would have the function of disclosing the responsibilities of the political actors within the IPCC. It would hinder the use of scientific argumentation to camouflage what are actually political decisions. Furthermore, it would help unencumber science from value questions, which are solely within the decision-making realm of political actors and the political system.

The unique contribution of the IPCC would be the organisation of the conception and writing of all three reports, and then linking them to one another. Report 1 should deal exclusively with scientific facts, without any consideration of political aspects. Only the IPCC scientists involved would be responsible for the contents. Reports 2 and 3 must not argue with facts that are beyond verification in Report 1. Report 3 must not introduce scientific arguments or political evaluations and allies beyond the road map of alternatives presented in Report 2. Report 3 is free to select the preferable options and should not be criticised for the selection by the scientists or the political actors within the IPCC. Of course, the IPCC speaks on behalf of its members only, so that everybody outside the IPCC is welcome to make scientifically-based or politically motivated criticisms (Fig. 5.1).

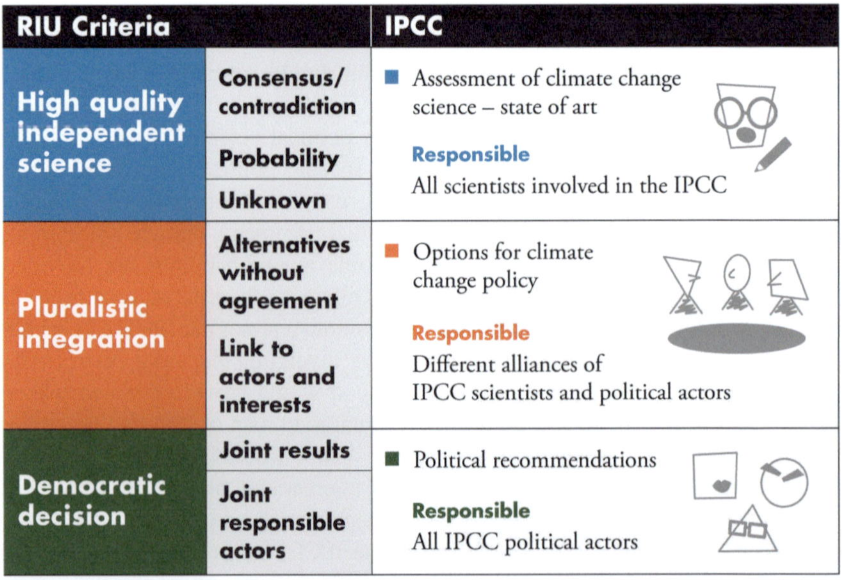

Fig. 5.1 Improvement of the IPCC according to RIU

5.3 Optimising the Potential of Science and Politics for an Effective and Efficient German Advisory Council on the Environment

The German federal government established the German Advisory Council on the Environment (SRU, *Sachverständigenrat für Umweltfragen*) in 1971 as part of the German environmental programme. The SRU serves as a scientific advisory body for environmental policy-making in Germany. Since its foundation, the SRU has delivered several general reports dealing with the overall quality of German environment and German environmental policy, and it has provided suggestions for environmental policy measures and described environmental problems that should be solved by politics in future. In addition to these "big" general reports, the SRU has also delivered reports on specific environmental policy issues, like waste, nature conservation, or rural development. Currently, the SRU consists of seven council members, representing different environmentally relevant scientific disciplines. The German Ministry for the Environment (BMUB) appoints these members for a four-year period, based on a cabinet decision. All members are university professors and can be re-appointed. The environment council is assisted by a permanent secretariat.[1] The German SRU is a well-established, long-term important actor in scientific environmental policy consultancy in Germany.

For 45 years, the German SRU has provided scientific expertise in environmental issues for political actors, society, and the economy. The fact that the SRU has been frequently and heavily criticised by many actors is not a bad sign for such a representative voice in environmental issues. Even correct scientific advice for the purpose of strengthening environmental protection will not please everyone in politics all the time. But a critical self-evaluation by the SRU reveals some deficits, and also solutions to improve the SRU and its political impact (Hey 2009), in addition to its many achievements. One problem of the SRU's work in the past was the production of "big reports" every 2 years and, since 2005, every 4 years. This seems to be ineffective because these big reports are not read by political actors in their entirety and cannot always deliver scientific suggestions at the time they are needed. The main idea to solve this problem is to differentiate the products of the SRU so as better to carry out specific tasks (Hey 2009, p. 261). The time period between the deliveries of the main reports was prolonged in order to create more space to work on specific, issue-oriented, smaller, but more timely reports. Currently, the SRU publishes the "Environmental Report" every 4 years, "Special Reports" on more specific environmental issues, and shorter "Statements" and "Comments" on certain environment-related policy debates or regulations.[2] The central "environmental report" should focus on scientific reporting and evaluation,

---

[1]http://www.umweltrat.de/EN/TheGermanAdvisoryCouncilOnTheEnvironment/Council/mission_node.html.

[2]http://www.umweltrat.de/EN/Reports/reports_node.html.

whereas the special reports can stress the announcement of new environmental challenges. The shorter statements and comments are the best way to deal with contradictory issues and to provide scientifically based estimations. The overall aim is to strengthen science-based environmental policy in relation to interest-driven politics within the political system in Germany (Hey 2009).

The RIU model is able to specify the necessary differentiation of the SRU's products further and to draw attention to factors contributing to successful scientific knowledge transfer.

1. Traditionally, the central "Environmental Report" combines science-based reporting with evaluation. The evaluation is driven by public values. Historically, the value-orientation shifted significantly between an economic and an environmental focus, depending on the selection of the seven members of the SRU by the German government. The evident dependence on the particular personality of the scientific members weakens the scientific credibility of the whole report, since it reflects scientific approaches and thematic focuses of the seven council members, but not necessarily the most current knowledge in environmental science as a whole.

Seen through the lens of the RIU model, a decline in the scientific quality of the report is an inevitable consequence of mingling science-based reporting and value-laden evaluation. In addition, the RIU model does not assume, as a given, that the joint evaluation by seven SRU scientists will have a strong political voice, in spite of their having a public mandate via appointment from the German government. The SRU lacks the necessary power sources to have its message weigh against the interests of political actors. However, lack of political power does not necessarily hurt the credibility and the scientific importance of a purely scientific report on the environment. Such a report would only have to stand against the scrutiny of reviews within the scientific discourse to assert its relevance.

The purely scientific environmental report would deal mainly with the most recent results on selected, important issues, in the environmental sciences. It would show areas of scientific consensus, but it would also underscore contradictions in results from research in environmental science. In addition, it would provide access to the statistical data revealing the probabilities associated with such results and would reveal facts previously unknown to science. The report has to be closely linked with the scientific discourse, which will complement the scientific competence of the seven members of the SRU, which would be unavoidably very limited. Nevertheless, the seven members are responsible for checking all statements thoroughly for soundness, in the context of the scientific discourse. The report would have great value for the environmental research community in Germany as a whole, as well as at an international level, and it would provide the best scientific basis for scientific knowledge transfer to politics.

2. The evaluation portion of the traditional "environmental report" could be separated from the scientific report and could be designed according to the RIU model's recommendations for pluralistic integration. "Political options for specific environmental issues" could be produced by active scientists designing alliances with specific actors in politics, economy, or society who—driven by their interests—

might be willing to use their political means to promote a specific option. Designing multiple alternatives for different alliances would open the decision space for science-based evaluation and advice. As the historical experience of the SRU shows, the members consider the interests of the government and political parties in opposition in any case, when they pursue evaluations and draft recommendations. Their aim is to come up with a message that is strong ("robust") enough to be viable in the political discourse (Hey 2009, p. 263 f.). The SRU has never come up with recommendations that deviate fundamentally from the political environmental discourse in Germany, even when there are good scientific arguments to rethink everything anew.

In the RIU model, integration needs an orientation toward public goals. The evaluation and recommendations gain relevance when one establishes explicit links to the actors who are in favour of specific public goals. This fosters political support for the alternatives. Of course, not all actors will gather behind a specific solution. The bias can be minimised overall by means of a pluralistic integration that delineates alternative solutions, ones for which the scientists then win different allies.

The existence of multiple alternatives and of alliances of supportive actors will open the opportunity for the SRU recommendations to become more specific. The pressure to cover all interests within one well balanced recommendation decreases and the danger of extremely general formulation is avoided. Instead of keeping the recommendations vague in order to avoid critique by powerful political actors, e.g., the government, the recommendations become specific and therefore helpful in making science-based options in environmental politics visible.

One important role of the SRU is to make decisions concerning highly-contested scientific alternatives in environmental issues. The RIU model recommends specificity regarding contested alternatives but also the restriction of the SRU contribution to the determination of whether and which of the alternatives are founded on a sound scientific basis. The SRU can carry out this scientific evaluation in a manner that is backed up by its scientific competence. A further, political, evaluation of alternatives should be avoided because it would involve judgement of alternatives based on the values of the members of the SRU. Such value judgements by experts do not have much significance within the multiple values of a pluralistic society. Linking value judgments to its scientific expertise would weaken the scientific authority and credibility of the SRU.

The group of seven members does not share responsibility for every alternative jointly. Instead, individual members are in charge of different alternatives, together with the scientists they rely on and the political actors they address regarding the specific alternative they are proposing. The task of the SRU as a whole is to guarantee that the individual responsibilities are made visible.

3. Finally, the RIU model indicates clearly that political decisions have to be made solely by the political actors. The SRU lacks the power to make relevant

political decisions on environmental issues. Its high scientific competence does not translate into competence in political decision-making. Political decisions imply making choices between different values and public goals and need to be backed by power resources.

The RIU model assumes that scientific recommendations that avoid value-based decisions and that target specific political allies will have a much greater impact in practice than value-laden statements. Specific allies will support selected scientific recommendations and implement them even against the resistance of other political actors driven by other interests. The SRU would gain influence by focusing on scientific arguments and letting the political actors fulfil their political tasks.

From the point of view of responsibility, the experts of the SRU lack a strong democratic legitimation to make political decisions. The SRU has a legal mandate to perform evaluations that goes beyond the purely scientific analysis of the state of the environment (Hey 2009, p. 163). However, due to the limitations of science in the type of value-based decision-making required in evaluation, the SRU's decisions would gain more political weight if they were restricted to the design of alternatives linked to specific allies, than if they involved a vote for a "First-Best Option" (Hey 2009, p. 257). In addition, this restriction would force political actors to bear their responsibility. Environmental sciences and democratic politics would benefit if there were a clear division of responsibilities. Science-based environmental policy would be strengthened (Fig. 5.2).

RIU Criteria		SRU
Independent science	Consensus/ contradiction	■ Report on environment – state of art **Responsible** All SRU scientists
	Probability	
	Unknown	
Pluralistic integration	Alternatives without agreement	■ Political options in specific environmental issues **Responsible** Selected alliances of scientists and political actors
	Link to actors and interests	
Democratic decision	Joint results	■ No political recommendation by SRU of "First-Best-Option" **Responsible** German democratic institutions
	Joint responsible actors	

Fig. 5.2 Improvement of the SRU according to RIU

5.4 Optimizing a Realistic Phronetic Social Science

The "phronetic social science approach" is one prominent example of how to engage in social science in order for it to become relevant in life (Schram 2012, p. 15). "Real social science" focuses on enhancing practical wisdom which promotes positive social change, instead of engaging in futile scientific attempts at explaining social life by means of an "abstract knowledge of law-like processes" and instead of using social technology, which will not work anyway (Schram 2012, p. 20).

The need and options for phronetic social science are described by referring to problems for which mainstream social sciences have to become relevant if they are to effect needed reforms in society and policy, and by fundamental ontological and epistemological arguments that show why social sciences cannot be practiced successfully by simply following the basic principles of natural sciences.

It is no wonder that the general approach of phronetic social science (Flyvbjerg 2001) is heavily discussed and partly criticised by mainstream social science (e.g. Laitin 2003; Falk et al. 2009). The RIU model is not a proper instrument to join this meta-scientific discourse. It works with all kinds of scientific research and shows how the political process dominates the transfer of scientific knowledge into practice. Therefore, we leave the ontological and epistemological discussion open and instead of this, we focus on the process by which scientific knowledge that is produced in a phronetic manner can become relevant in practice.

Looking at phronetic social science through the lens of the RIU model reveals key factors for the transfer of scientific knowledge that the phronetic approach looks at as well. It explicitly reminds the researcher to look at the power of actors and at the benefits and losses they face by remaining within the status quo, in contrast to those they will see by implementing innovative scientific solutions (Schram 2012, p. 20). Furthermore, it stresses that the questions of what is desirable and what should be done become important if scientific knowledge is to be applied. The aim of phronetic social science is very close to that of the RIU model, and the main factors in both worlds are similar. Nevertheless, the RIU model could support the ways in which the efforts of phronetic researchers may become relevant in politics and may help establish where to draw political limits.

Phronetic social science "focuses its efforts on answering four critical questions related to enhancing practice wisdom: (1) where are we going?; (2) who gains, and who loses, by which mechanism of power?; (3) is it desirable?; (4) what should be done?" (Schram 2012, p. 19 f.).

1. Science is well prepared to answer the first two questions if sound methods are applied. Analysing the consequences of existing policies, evaluating winners and losers and explaining the results by means of power processes is the very competence of social science based on empirical research methods and empirical facts. Most methodological concepts focus on fulfilling these tasks scientifically at a high standard. The concept of "real social science" elaborates, in addition, that sound methodological standards have to be applied (Flyvbjerg et al. 2012). The RIU

model warns that the interventions of political actors endanger high standards within applied science. This danger is especially high for phronetic researchers who communicate closely with political actors in the course of the entire research process. Some interested actors will strongly pressure the scientists that their statements about the consequences and distribution of benefits fit into actors' interests. They will not accept critical results if these may pose a disadvantage for them in the political process. The process of influencing scientists is highly complex and often hidden. For example, values shared by scientists and actors in practice may keep phronetic scientists from accepting and publishing results that endanger the position or power of specific actors.

The RIU model offers a way to avoid this threat against good scientific practice, by distinguishing research from integration. Professional integration guarantees at best that the scientific results will be linked to the needs of actors. But the concept of independent science protects researchers and enables them to apply scientific methods at a high standard and independently from any interventions. The RIU model warns phronetic researchers not to underestimate the power of actors to intervene in the research process. Driven by their interests, such political actors simply do not accept researchers' specific scientific or normative arguments. From their interest-oriented point of view, some scientific arguments seem meaningless, even despite the fact that they are empirically proven.

Phronetic social science is confronted with the danger of political intervention because researchers exchange information with actors in practice, and all actors follow their specific interests. The RIU model recommends securing a protected exclusive space for the process of phronetic social science, where sound methods can be applied rigorously and results are accepted independently from whether they may benefit someone in particular. The responsibility for keeping high methodological standards is a task exclusively for scientists and their independent scientific discourses.

2. Of course, phronetic social science does not stop with the scientific results. It links them closely to the answers of the third and fourth questions, of what is desirable and what should be done. The very heart of politics is that there are always multiple desirable results and many courses of action. If a first-best option existed in each case, politics would not be necessary to help decide what to do. The RIU model offers a solution to this dilemma of scientific advice on a first-best option by means of pluralistic integration.

Keeping in mind that each solution will always benefit specific actors more than the alternatives and will be biased toward specific public goals, the RIU model recommends that the link to public goals and the interests of specific allies of actors should be made visible. This can be done by showing different options within the specific context.

Each option is based soundly on the scientific results. Additionally, the options are linked to the interests of selected actors. To support a preferable option, scientists should look for allies actively. Since the phronetic approach focuses on disadvantaged political actors, it is most interesting to look for new alliances between weak, disadvantaged actors and new, strong actors. The bad news is that designing an option solely for disadvantaged actors will not do much for political change toward a better quality of life, as long as no powerful actor can be found as an ally. The good news is that within a pluralistic society and politics there are multiple options for alliances with different actors. Making visible the links of specific options to specific actors is the best way to engage them and to optimise the different areas of expertise of scientists and political actors. The selected scientists and the political actors who stand together behind a given option would be jointly responsible for the recommended changes.

3. Whereas selected scientists and political actors would be able to achieve consensus on specific options, it is unrealistic to assume that society or politics as a whole will achieve a consensus in any relevant political issue. What politics can do is to decide between the options and to assert the decision using power-based means. The key idea behind democratic political decision-making is not that there is a consensus on specific options, but rather, that there is consensus on the notion that decision-making will be organised in a democratic manner.

The RIU model, which is based on political science theory, draws attention to the very limited power of scientists. They have almost no means to make a difference within the power play of politics. Therefore, scientific results become politically relevant not because of interventions by scientists arguing how important those are, be but because of political allies who gather around specific scientific results and push scientific rationality through against the resistance of other political actors by means of their stronger power. Within this power struggle in day-to-day politics, scientific discourse has little significance because other power means, like economic resources, legal rights or technical resources, are stronger drivers for decision-making.

The power process of politics means that phronetic researchers should realise the limits of scientific influence. Not much empirical evidence can be found in practice for support, in the media or public discourse, for scientific, sound solutions, however rational arguments may be. The media are also driven by strong economic interests, and the public discourse is not able to mirror complex and judicious arguments correctly in the long run. The new media have not changed the public preference for "sexy" news and the priority for the very latest events, thus burying long-term sensible discourse often before it can ever become relevant.

The RIU model recommends that phronetic social science should be realistic about the small political power it has to influence binding decisions. Presenting selected, solid, science-based options linked to actors is the major form of influence phronetic social science can use. Speaking truth within power is a strong achievement of phronetic social science. Attempting to speak truth to power so as to have a direct impact on interests and power-driven decisions is futile, however. The political weight of a phronetic researcher does not differ much from that of a single citizen. It may be sobering for researchers to realise the dominance of political actors' strength. However, this dominance implies that phronetic social science cannot be held responsible for political decisions, and that elected actors must bear that responsibility, something that is advantageous for the democratic order. Furthermore, this division of responsibility, where phronetic social science is

RIU Criteria		Phronetic
Independent science	Consensus/ contradiction	■ Protected space for scientific methods **Responsible** Scientists 
	Probability	
	Unknown	
Pluralistic integration	Alternatives without agreement	■ Political options within context  **Responsible** Scientists and selected political actors
	Link to actors and interests	
Democratic decision	Joint results	■ Political recommendation only as citizen  **Responsible** As citizen
	Joint responsible actors	

Fig. 5.3 Improvement of phronetic social science according to RIU

responsible for high quality research and political actors for political decisions, can represent an advantage for phronetic social science researchers, as it helps them retain their credibility and avoid frustration associated with their relevance in politics, or lack thereof (Fig. 5.3).

5.5 Quality Management for the Network of the European Forest Institute (EFI)

The European Forest Institute (EFI) has developed an outstanding network of forest sciences in Europe during the last 20 years (European Forest Institute 2013). It is an international organisation established by European states. The mission of the European Forest Institute is to "undertake research…in order to promote the conservation and sustainable management of forests in Europe" (European Forest Institute 2003). It conducts research and provides policy advice on forest-related issues. In addition, it advocates for the further development of forest research.

The European Forest Institute (EFI) has established a strong network of 25 full member countries and 120 organisations from 35 countries as associate and affiliate members (European Forest Institute n.d., p. 6). Its projects deal with the important forestry issues in Europe and it provides science-based information on these issues.

The research unit is the backbone of the institute. It conducts interdisciplinary and cross-sector research mainly by networking the research capacities of its members. The EFI is a strong and successful competitor in the areas covered within its European research thematic programmes: "Sustainability and Climate Change"[3] and "Forest for Society".[4]

The main target group for EFI's policy advice is the European Union. The unit "ThinkForest"[5] organises a European high-level forum on the future of forests and offers seminars by scientists and stakeholders on emerging forest issues. With its "EU FLEGT Facility"[6] and its "EU REDD Facility"[7] the European Forest Institute supports the implementation of two central programmes of international forest policy against illegal logging and for the reduction of emissions from deforestation and forest degradation. The most recent unit of the EFI fosters a "Foresight and Policy Support Programme".[8] It aims to assist European political decision-makers, policy institutions and research organisations by means of science-based information.

The European Forest Institute is well established within the forest sciences and criticisms of its mission and practice are seldom heard. But this success story has its limits, and science-based policy advice has not yet become a strong feature for the transfer of scientific knowledge to forest policy, neither at a European nor at a global level. The mission of the European Forest Institute is still to be accomplished, and major challenges lay ahead. Therefore it makes sense to look for options to strengthen the Institute. The RIU model identifies important success factors for such improvements.

1. The research network of the European Forest Institute has gained much strength due to the sheer size of its group of 300 researchers, and also because of its outreach to the diversity of scientific disciplines that deal with forests. Additionally, the regional research facilities and members' competences shape the European Forest Institute in a unique manner: The EFI is built on richness of experience. By mobilising these competences, the Institute aims to carry out interdisciplinary, international and applied projects. Here lies the challenge for the future. Having a huge number of resources and setting challenging goals do not guarantee high quality research. Quite the opposite is true, because the vastness of the resources within joint projects and the demands of interdisciplinarity and internationality make it more difficult to meet high scientific standards. In practice, it is very difficult to coordinate the multiple theoretical and methodological approaches of different disciplines as practised by many researchers and to bundle the empirical data they produce in meaningful ways. The scientific result is often not more than a report of

---

[3]http://www.efi.int/portal/research/research_programmes/sustainability_and_climate_change/.

[4]http://www.efi.int/portal/research/research_programmes/forest_for_society/.

[5]http://www.efi.int/portal/policy_advice/thinkforest/.

[6]http://www.efi.int/portal/about_efi/structure/flegt/.

[7]http://www.efi.int/portal/about_efi/structure/eu_redd/.

[8]http://www.efi.int/portal/policy_advice/fps/.

many pages that paints a colourful picture that is, however, diffuse and very inconsistent in quality. The largeness and complexity of the projects threaten the scientific quality of the EFI's work. Another threat is the close link to actors in practice. Most projects pursue applied science goals, which means that they seek to contribute to problem solving in practice. A transdisciplinary involvement of stakeholders is often part of these projects. Additionally, the second main goal of the institute, namely, policy advice, strengthens the link to practice within each project.

It is a big challenge to maintain scientific standards against the strong demands of practice within the institute. The RIU model shows the decisive factors that protect the necessary independence of scientific research.

The main strategy is to link the research of the institute with the scientific discourse. This has a dimension of content and a dimension of organisation. Within the scientific content of each project, the European Forest Institute should make visible the potential for consensus but also the contradictions and uncertainties within the scientific state of the art. A scientific discussion beyond consensus is necessary to meet good standards of scientific practice, even if the complexity of a scientific report based on this meant that it could not provide meaningful advice to actors in the field.

There is unique scientific potential in the multiple languages and traditions of the members of the Institute. All forest issues have a local dimension rooted in the specific ecological and social context of the corresponding forest areas. This local aspect is neglected by the dominant English-speaking international scientific discourse. The network of the Institute offers a unique chance to link local contexts to general aspects within the scientific analysis in innovative ways in the future.

At the organisational level, there is a good option for improving the links of the Institute's research to the organisations of forest-related sciences. This goes far beyond the traditional forest sciences as represented by the departments of forestry at universities and by research institutes in Europe. There are many highly competent research organisations, e.g., in the disciplines of biology, geography, genetics or soil sciences, which could take part in the scientific discourse. Such interaction cannot be implemented in general, but each project has potential for a specific outreach to an area of scientific discourse. This would have strong positive effects on scientific quality.

The European Forest Institute has taken important steps to strengthen the scientific quality of its work. Among these are the establishment of the International Scientific Board and successful close cooperation with selected universities, especially in Finland. The RIU model underscores that strong research is a precondition for good policy advice and that the Institute as a whole bears the responsibility for the quality and independence of its scientific research.

2. Policy advice is one of the main tasks of the European Forest Institute (European Forest Institute n.d.). Multiple features like ThinkForest, Facilities for FLEGT and REDD and the Foresight and Policy Support Programme deliver

science-based information for political actors.[9] Effective advice cannot be produced by science alone. In addition, pluralistic integration incorporating the needs of the political actors is necessary. The RIU model also mentions success factors for integration that could be valuable for the optimisation of the Institute's knowledge transfer.

The Institute aims to ask and answer the question of the preferred future for the European forest sector. This key issue makes the challenges of policy advice visible. Is there just one, best option for the future and who would prefer it: the forest actors, the environmental actors or the public? If it is the public, which specific public, from the northern or the southern states of Europe? In the pluralistic political and social system of Europe there will be no consensus in answering such relevant questions about the protection and use of European forests. Therefore, an attempt to design a single, best policy advice for the European Union's forest policy will either fail or produce a general statement that offers no relevant information for the political actors. Everybody will accept a statement supporting sustainable forestry, but this general information will offer specific, and therefore relevant, advice to no one.

The RIU model solves this key dilemma associated with all policy advice by using a pluralistic approach that recommends the design of alternatives for a preferred future and to link these to specific actors. The European Forest Institute has close links to the entire forest sector, including the forest industry. The advice the institute is working on is science-based, oriented toward public goals and linked to the interests of strong forest actors. This means that the consensus with environmental interests or different public arenas cannot be achieved while this work is being done. A bias toward specific actors and interests is unavoidable in policy advice. The RIU model recommends to make this bias visible and to offer alternatives, in close cooperation with selected political actors.

It would be a too broad and therefore unfulfillable mission for the European Forest Institute to design policy alternatives for all actors interested in forests, ranging from the forest industry to NGOs like Greenpeace. But the forest sector offers some options for thinking in alternatives focused on specific actors. For example, in forestry issues there may be a northern consensus and a different, southern consensus. Industry may find some specific alternatives for the future to be preferable, and forest owners may prefer others. By diversifying policy advice, which is then meant for numerous, specific, individual political actors, it will become very relevant for these actors, and they will use their political resources for serving as backbone of the advice, giving it weight within the practice of forest-policy making.

The responsibility for alternatives is kept by the European Forest Institute and the specific political allies behind the alternatives.

---

[9]Information about EFI's different policy support programmes can be retrieved from http://www.efi.int/portal/policy_advice/fps/.

3. The European Forest Institute gets close to political decisions by being involved in the implementation of political programs like FLEGT and REDD and by lobbying for forest sciences. The policy support for FLEGT and REDD received a mandate from the European Union. According to the RIU model, such a mandate is important because science alone cannot make political decisions in forestry issues.

In forestry issues, the mandate of the European Union itself is rather general and weak because the treaty that established the European Economic Community (TEEC) did not mention forestry as a supranational task. But the European Union became active in environmental and regional policies that had an extensive impact on forestry, by means of non-binding forest governance (Krott 2008; Pülzl et al. 2013). The general and weak forest mandate of the European Union opened up decision-making space for the policy support of the European Forest Institute. The RIU model recommends not to overestimate the weight of scientific advice for political decision-making. In conflicting issues the political actors will judge science as having little weight, due to a lack of power resources that scientists can mobilise. Purely scientific advice will be simply ignored by decision makers. The policy support from the European Forest Institute becomes relevant for decisions only as far as political actors support it and only if they assume the responsibility for decisions based on it. Therefore, policy support from the Institute needs a specified mandate from the European Union and it will be successful in influencing decisions only if it stays within the boundaries of the mandate and if it is backed by the European Union. The same would be true for policy support based on a mandate from other actors.

The only political task for which the European Forest Institute has its own mandate is advocating and developing forest sciences. Being a network of scientific organisations and scientists provides democratic legitimation to the Institute's representation of the interests of forest sciences.

The dynamic growth of science at the European level creates a strong need for advocating forest sciences. Science policy between the European Union and the national member states is highly competitive (Geuna 2001). Only strong organisations of scientists can exert an influence against the technocratic concepts of European and national bureaucrats (Banchoff 2002). So far, there is no strong, unified reasonable voice in the forest sciences. The European Forest Institute's advocacy fills an important niche. The RIU model differentiates this advocacy for science from scientific advice clearly. In making science policy the European Forest Institute becomes a political actor that assumes responsibilities in full. Successful policy making will require the use of all political means exceeding scientific arguments by far (Fig. 5.4).

RIU Criteria		European Forest Institute EFI
Independent science	Consensus/ contradiction	■ Quality management for science **Responsible** EFI 
	Probability	
	Unknown	
Pluralistic integration	Alternatives without agreement	■ Alternatives for specific forest actors **Responsible** EFI and allies 
	Link to actors and interests	
Democratic decision	Joint results	■ Mandate from actors, e.g. European Union for policy advice **Responsible** European Union 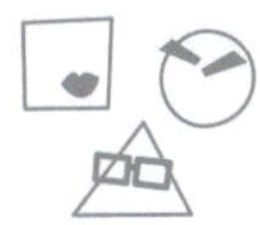 ■ Own mandate for advocating science **Responsible** EFI

Fig. 5.4 Improvement of the EFI according to RIU

5.6 Quality Assurance of the German Federal Agency for the Environment (UBA)

The German Federal Agency for the *Environment* (*Umweltbundesamt*, UBA) is a departmental research institution founded in 1974 (Böcher and Töller 2012). The UBA is Germany's central federal authority on environmental matters and it is the largest of three subordinate authorities of the German Environmental Ministry (*Bundesministerium für Umwelt, Naturschutz, Bau und Reaktorsicherheit*, BMUB), with a staff of more than 1200 employees (Böcher 2010). The UBA's mission is legitimised and defined by public law (*Errichtungsgesetz*) and consists of providing science-based advice to the federal government, administering environmental legislation, and providing relevant environmental information to the public (Rose-Ackerman 1995). The three most important general objectives of the UBA are (1) to provide scientific assistance to the BMUB and other federal ministries on matters concerning environmental protection, including its health implications; (2) to apply important legislation involving expert scientific knowledge; and (3) to inform the public about environmental protection. To fulfil these tasks, the UBA conducts independent research (performed by the UBA's own scientific staff) or commissions research projects to be conducted by external institutions, such as universities or private consultancies. The UBA analyses and interprets both its own and commissioned research and transfers the science-based policy knowledge to the relevant target groups (ministry administrators, the public, environmental non-governmental organizations, the media, and others) via methods such as internal reports (for the Ministry), scientific or consumer publications, the Internet, brochures, and press releases. All research and administrative tasks undertaken must contribute to the three main objectives of the UBA. This requirement marks an important difference between the UBA and other German research institutions, such as universities, which are largely able to define their own research objectives.[10] The definition of projects to be commissioned by external research institutions must be coordinated annually with the BMUB, which must approve the research programme.

Since its foundation 40 years ago, the UBA has become a well-respected authority regarding environmental policy questions. Nevertheless, in the recent past there was a public debate in Germany surrounding the quality assurance of departmental research institutions in general (Wissenschaftsrat 2007; Barlösius 2010), leading to an evaluation of all German departmental research institutions by the German Council of Science and Humanities (*Wissenschaftsrat*) that was finished in 2010 (Wissenschaftsrat 2010). One of the main controversial topics concerning the role of departmental research institutions and the UBA was how the quality of independent scientific research can be guaranteed despite its being part of the government as a subordinate authority of the Environmental Ministry. In addition, sociologists, in particular, claimed that agencies like the UBA have to deal with doing research, but political expectations have to be fulfilled by delivering science-based products that reflect the interests and demands of political actors. (Barlösius 2009). The subsequent argument was that, due to its increasing administrational and political tasks, the scientific quality of the UBA's research necessarily had to decline. The main challenge for the UBA's quality assurance would then be in balancing political and scientific tasks whilst avoiding loss of quality in each respective subtask. This public debate reflects very well the main challenges for scientific knowledge transfer as described in the first two chapters of this book: Politics and science are not comparable and therefore a hybrid institution that represents modes of thought from both, like the UBA, cannot work properly without developing deficits in one of these two different spheres of activity (Barlösius 2009; Döhler 2007; Hohn and Schimank 1990).

According to our RIU model, quality assurance of the UBA would consist of a clear separation of three UBA-subtasks: The UBA has to conduct research (on its own or by commissioning extramural research projects), it has to be active in carrying out professional integration, and it has to deliver science-based products for utilization in practice. The UBA's activities are often initiated by the BMUB within the sphere of integration. A political demand for the science-based advice of the UBA leads to projects in the sphere of research, for which the results again lead back

[10]Nevertheless, German universities have become increasingly dependent on third-party funding.

to the sphere of integration. This process is continuous, until the final science-based advisory product is delivered to the external sphere of utilization. Within the utilization sphere, employees of the UBA can no longer change the advisory products —such products lie in the hands of various political actors who decide how to use them. This alternation between research and integration illustrates how individual products of the utilization sphere always incorporate components of research and integration. By using the production lines of the RIU model it is possible to make evident the time resources that are invested in the different activities.

The research tasks of the Federal Environmental Agency cannot consist solely of conducting basic research. All research projects must answer practice-oriented questions that are raised within the sphere of integration. This is due to legal aspects of the "*Errichtungsgesetz*" of the UBA. However, in order to remain true to the ideals of independent science, the Federal Environmental Agency must guarantee that its research represents the "state of the art" of those scientific theories, methods, and data that were previously chosen during a process of integration. Furthermore, it is very important that the UBA should assess all relevant scientific sources regarding specific problems and that there should be no censorship due to political opportunities. Furthermore, existing consensus, contradictions, and probabilities regarding scientific prognoses about future developments have to be expressed clearly. As a result of independent scientific research, the UBA has to provide all existing research results and their consequences for political decisions to the Ministry, without altering scientific findings and without interpreting them in advance. In many environmental policy issues, the UBA might deliver alternative political options that are the consequence of different scientific discussions that lead to different science-based consequences. It is important that the UBA should reveal scientific controversies. Another important resulting possibility if the UBA's independence in science practice is to be secure, is that it may deliver science-based integration results that are against the interests of the Ministry at that given moment. In the UBA's history there have been many cases in which the UBA provided science-based information and recommendations even against the political interests of the current German government. In the 1990s the UBA voted for a German environmental tax reform long before the German government changed and such a reform was politically adopted. Another example is the UBA's vote for a maximum speed limit on German highways, which is not politically feasible, regardless of which government is in office. This freedom the UBA has, of providing science-based information without having to anticipate possible political reactions is one very important aspect of quality assurance by using the RIU model. Of course, it is possible for political actors to neglect or disregard the recommendations of the UBA—but that is due to political rationality. Therefore it is for political actors to assume responsibility for this, and the resulting possibility of bad decision-making, and not for the scientific staff of the UBA. To meet such criteria of independent science practice, the research activities of the UBA incorporate the most current research on specific topics, initiate and steer external research projects, conduct internal research by performing laboratory experiments or round robin tests, develop methods, maintain a scientific research desk for answering short-term inquiries, participate in (inter)national scientific conferences, and publish scientific materials, among other tasks (Böcher and Krott 2010, p. 32 f.).

The Federal Environmental Agency provides pluralistic integration, as its staff creates links between scientific research and practical problems, and vice versa. This task is typical for departmental research institutions, and it can be characterised as the institutionalization of integration activities. As noted earlier in this book, integration refers to a bi-directional process between practical utilization and research. Integration activities are conducted by the UBA's employees, who are scientifically trained and who have substantial experience in working at the intersection between science and practice. In the course of the external research projects commissioned by the UBA, these employees are responsible for interpreting research results concerning practical or political demands, or for modifying research projects in the event that they lose their practical relevance or that they must answer research questions that are not relevant to politics. Long-term experience in cooperation with the ministries and other authorities is important for the integration role of the UBA. Furthermore, knowledge regarding the interests of the industry, environmental NGOs, and citizens is crucial for pluralistic integration measures. In addition, it is important to show alternatives that reflect different scientific results and to link these alternatives to different actors and interests. To assure the quality of integration, the UBA has to consider the resources of different actors as potential allies for the implementation of the UBA's science-based advice, and it has to provide results of its work in the form of target-group intermediation. The UBA's integration tasks incorporate the following actions (Böcher 2012, p. 471; Böcher and Krott 2010, p. 36): the formulation, assessment, and selection of scientific research questions based on their potential contribution to the solution of problems identified by the ministries; the formulation of departmental research questions to initiate research projects; the writing of statements for ministries or EU bodies; the writing of statements about research results and political implications for the ministries; and the formulation of answers to inquiries from citizens and from other political actors. It is important that the UBA's research and integration activities are not always separate from one another. There is a narrow connection between these two activities, and integration often occurs within the minds of the employees. All of these integration tasks are based on research and add practical actor-oriented considerations that lead to different utilization products.

The main difference between utilization, research, and integration is that, within the utilization sphere, the UBA comes into contact with certain target groups to deliver its products. Utilization refers to activities that address external actors and change their state of information. Finalised science-based advisory products are delivered to political actors, and the UBA staff cannot change them further. Nevertheless, the active use of the science-based information from the UBA, and its consequences, are the joint result of the UBA's research and integration activities and their use by different political actors. Typical utilization products of the UBA may include recommendations to citizens via brochures (e.g., "*Hilfe! Schimmel im Haushalt*," which translates as "Help! Mold at home!" (Umweltbundesamt 2005), regulation information, the delivery of implementation guidelines and scientific methods, and the delivery of products for the scientific community (e.g., publications). It is important that the decision about what science-based information generated by the UBA is used in political practice should remain in the political sphere, as part of democratic decision-making. The UBA itself has no political mandate to make or enforce such decisions, which must be made by political actors who have to assume the corresponding responsibilities with regard to their constituencies.

The RIU model delivers criteria for an effective quality assurance of the UBA. The model reflects the continuously changing activities in the research, integration, and utilization spheres that constitute the groundwork for the UBA's science-based policy advice. These activities can be described and analysed using the RIU model. Research, integration, and utilization are the foundations of the UBA's scientific advice. The UBA, as a departmental research institution, must address both scientific requirements and practical demands simultaneously. Nevertheless, it is important for the sphere of research to guarantee scientific independence, for integration to have a pluralistic orientation toward public goals, and for utilization to restrict science-based political implementation to the specific area for which the UBA receives a specific mandate from political actors. In general, the UBA should not try to make political decisions; it should leave this duty to the political actors who are put in charge through a democratic process (Fig. 5.5).

RIU Criteria		UBA
Independent science	Consensus/ contradiction Probability Unknown	■ Quality management for science **Responsible** UBA 
Pluralistic integration	Alternatives without agreement Link to actors and interests	■ Alternative science-based information for different actors **Responsible** UBA and BMUB 
Democratic decision	Joint results	■ Mandate from political actors for specific science-based implementation **Responsible** BMUB and others 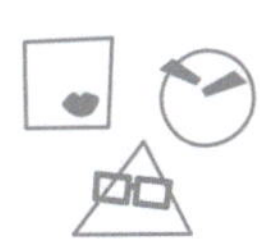
	Joint responsible actors	■ Own mandate for science-based recommendations **Responsible** UBA

Fig. 5.5 Improvement of the UBA according to RIU

5.7 Diversifying Responsibility for "Responsible Research and Innovation"

The recent dynamic developments in technology make evident the huge potential for innovation to serve societal and economic needs, e.g., the growing importance of health issues or the very dynamic transition to renewable energy systems. Nevertheless, serious problems arising from technological innovation have become evident as well. Important research and innovation efforts have led to contested technologies, like genetically modified organisms (GMOs), derivation of human stem cells, or the development of nuclear technology. Considerable innovation and investments have been made, but there has also been an emergence of social and political conflicts, with negative consequences for business, public policies, science and society. If new products do not become marketable or if they are stopped due to political moratoriums, great losses of private investments are inevitable. In addition, public innovation policies and science would lose credibility. The suspicion would spread that societal needs and ethical concerns are neglected due to economic arguments, endangering the societal and political support for innovation.

The European Union's "Responsible Research and Innovation" (RRI), a comprehensive approach to proceeding in research and innovation so as to "achieve a better alignment" of the two with societal needs, claims to provide a solution for these costly and risky failures of science, by fostering two points of reference in the design of research and innovation processes and products (European Commission 2013, p. 56; see for an overview of the concept Owen et al. 2012). First, "ethical acceptability" has to be achieved, which means compliance with both the EU Charter on Fundamental Rights as well as the safety standards for the products. Second, there has to be an orientation towards societal needs products (European Commission 2013, p. 56). The objectives of sustainable development, and other core objectives expressed in the Treaty of the European Union, should be met. RRI considers ethical aspects and societal needs. It should be (i) anticipatory, in that it looks at new options and risks, (ii) inclusive, by involving as many stakeholders as possible, (iii) reflexive, in that it is aware of its own ethical, political and social assumptions and (iv) responsive, by being able to change research and innovation processes according to public values (European Commission 2013, p. 57 f.). RRI shows great potential for avoiding solutions that may later have a high ecological or societal cost or may be unacceptable to society and thus terminated, causing great financial losses for private and public investors. RRI could foster sustainable development as long as it is implemented in practice. Here, a major challenge for the approach arises because most of its arguments are ethical in nature, and it suggests that stakeholders, citizens and the public should behave in a new, more responsible manner.

The RIU model can elucidate how the RRI approach influences the driving forces in research and practice and how the links can be used to diversify responsibility in order to implement the approach.

(1) The potential of science to inform about facts and probabilities associated with solutions and its consequences, as well as to identify the realm of the unknown, is vast but limited. RRI aims to create awareness of the consequences of research for ethics and societal needs. Responsibility requires decisions on these subjects. Most often the scientific knowledge is too limited to make such decisions, but nevertheless science provides valuable information to improve them. The scientific information is more valuable if the science in question is state-of-the-art. The scientific methods applied have to be sound and independent from the aspirations and needs of all actors and the public. Innovative and empirically demonstrable results can be expected only if such independence is provided. This requirement of high quality science also applies to the implementation of the "Responsible Research and Implementation" approach. If stakeholders are allowed to intervene in current scientific processes or to vote on scientific facts the value that science contributes is endangered.

The RIU model recommends that the overall responsibility should be diversified and that the responsibility of science should focus on the state-of-the-art quality of facts and probabilities concerning solutions and consequences. Although science makes progress, many important consequences remain scientifically unknown at any given historical moment. The responsibility of science is always limited to the realm of what is scientifically known. It makes no sense to extend the responsibility of science beyond anything that is scientifically known. Science simply cannot provide scientifically sound answers regarding things that are beyond the scientific realm.

Within these limitations, science has a specific ability to produce sound information. Therefore, scientific organisations and scientists should only assume responsibility for producing scientific information that is state-of-the-art. Sharing this responsibility with other actors adds no value; on the contrary, it endangers the independence of science and, as a consequence, the quality of scientific enquiry. However, science can contribute what pertains to this specific and limited scope of responsibility generally to the processes of the "Responsible Research and Innovation" approach.

(2) According to the EU, "Responsible Research and Innovation" should open up science to the inclusion of as many stakeholders as possible (Geoghegan-Quinn 2012). Their experience and values are important to help identify societal needs and to build up responsibility. However, the incorporation of all these actors conflicts with the scientists' freedom of choice in the application of scientific methods that are based on defined procedures that leave no space for intervention by actors other than scientific actors. If value- and interest-driven actors have access to the scientific processes, the unavoidable consequence is that values and interests will influence, or even select, the production of scientific results. Stakeholders try to suppress undesired results, which diminishes strongly the critical and innovative potential of science. This important potential can only be realised by independent science.

The solution offered by the RIU model is to keep research independent, but to link it to practical utilization through a specific sphere of activity, called integration.

Within integration, the broad inclusion of actors fosters the selection of scientific results that are oriented toward ethical norms and societal needs. It is impossible to give a direct voice to society itself, but diverse actors can contribute their pluralistic views on responsible innovation products. These multiple actors will not achieve a consensus about what is right and responsible innovation, and in an open society nobody can force them to agree on one single solution. Therefore, a pluralistic approach is more promising. Here, different actors group around alternatives. The RIU model suggests that, within integration, scientists should leave their objective roles and should join a specific group that supports a specific solution. What is most important is that the members—scientific, economic and societal—admit openly that they support a specific solution and that they are willing to assume the responsibility for advancing this solution. Due to the diversity of societal needs, values and interests, a bundle of the many options that are each considered to be best will be advocated by different groups willing to assume this responsibility.

The task of pluralistic integration is to guarantee the space to bring forward alternatives, to provide possibilities for every actor to participate, and to make the supporting alliances of scientific, societal, economic and political actors visible. Scientific organisations alone would be overwhelmed by this task. Their competence does not suffice to deal with the contested aspirations within society and with any occurring intense conflicts. Successful scientific organisations have to be built on freedom for intrinsic curiosity about the world. This freedom for researchers is not compatible with an organisation that includes strong actors driven by interests in its research tasks.

The RIU model suggests new organisations for professional integration. The concept of "Responsible Research and Innovation" provides a golden opportunity to design and implement such a new type of organisation. Professional integration requires openness both towards science and practice. Professional integration is located between science and practice, which means that there are unique tasks that can be carried out by professional integrators within a dedicated integration organisation. Professional integrators play neither the role of scientists nor the role of actors in practice. They act between both spheres and manage the optimisation of bi-directional communication to enable a well-informed selection process. In this process, specific actors may select what is relevant and preferable for them from the collection of scientifically supported knowledge, and the scientists may select from the relevant practical problems and topics for scientific research. The organisation of integration that can be a task for integrators guarantees that only scientifically sound information is included and that as many actors as possible have free access to this information. This results in multiple alternatives for innovative solutions and visible allies of scientists and actors of the practice who are willing to carry responsibility for one or several specific solutions.

The challenge for the new integration organisation is to provide and protect such a free meeting space for researchers and actors engaged in practice. In practice, the free exchange of information is seriously endangered by various state and private actors who seek to dominate the discourse. To hinder dominance and to provide a fair inclusion of all participants, the integrating organisation has to be very independent and politically strong, and it should be legitimised by a specific law comparable to the existing public audit offices. What we suggest is a new "Clearing House Responsible for Research and Innovation".

In conclusion, professional integration would create a new protected space for the inclusion of all societal, political and economic actors, but without the obligation of producing a consensus. Such a protected space will contribute to a pluralistic society that can be a force against the strong dominant actors who usually drive the discourse, often into a biased pluralism dominated by most powerful (economic) actors over weak interests.

(3) Innovation can be realised after a decision about a specific innovative solution is made and implemented. The "Responsible Research and Innovation" approach claims that the solution should respect ethical norms and societal needs in an anticipatory, inclusive, reflexive and responsive manner. The practical challenge is what to do if no consensus can be reached on innovative solutions and the consequences this would have with regard to these requirements. The RIU model suggests that the solution would be that politicians who have a public mandate are the appropriate actors to decide responsibly.

Here, a political decision is responsible if it stays within the limits of the alternatives that the integration process offers. Integration guarantees a strong basis with regard to ethics, societal needs and scientific knowledge. But, in the final decision, one alternative must be selected, keeping in mind that many risks and future developments are still unknown. This task goes far beyond science and integration and is the sole duty of political actors who have a public mandate. RIU stresses that the role of political decision makers cannot be substituted by science or by an inclusion of societal and economic actors. In the measure in which private investors finance innovation, they share responsibility with the political actors. It is the duty of the state to set legally binding limits to the decision space and the responsibility of private investors.

Finally, political actors and investors also make decisions concerning the funding of research. Science lacks sufficient resources of its own to do relevant research, so it needs funding from society and the economy as a whole. The RIU model shows that "Responsible Research and Innnovation" requires independence from public and private intervention. A reasonable amount of funding has to be given to scientific organisations without any conditions that would restrict the topics or problems they may address. Within the basic ethical framework, science has to establish a space for theory-driven research, producing results that are always unforeseeable. Scientific responsibility means only to follow the principles that guarantee that its scientific research is state-of-the-art. Of course, actors have the option to orientate their funding of research toward solving specific problems. But RIU shows that such mission-oriented research can neither mobilise the whole innovative potential of research nor can it substitute pluralistic integration or democratic decision-making. An alternative would be to restructure programmes of mission-orientated research according to the RIU model.

RIU Criteria		Responsible Research and Innovation
Independent science	Consensus/ contradiction	■ State of art in facts and probabilities about solutions and consequences **Responsible** Researchers and research organizations
	Probability	
	Unknown	
Pluralistic integration	Alternatives without agreement	■ Alternatives in problem solutions **Responsible** Clearing house providing space for different alliances of scientific, societal, economic and political actors
	Link to actors and interests	
Democratic decision	Joint results	■ Decision in innovative solutions **Responsible** Investors and political organizations
	Joint responsible actors	

Fig. 5.6 Improvement of "Responsible Research and Innovation" according to RIU

In all, the RIU model recommends the diversification of the specific competences and responsibilities of (1) research, (2) integration within the economy and society and (3) political decision making and then to link them together in a new procedure so as to come up with the best version of a general "Responsible Research and Innovation" which works (Fig. 5.6).

5.8 Effective Science-Public Communication

Despite growing scientific progress and the ubiquity of science in public discourse and the media, one can observe a growing scepticism regarding science (Kraft et al. 2015). In a way that is not typical of our modern society, citizens are sceptical even in cases in which there is overwhelming scientific evidence: Many people doubt the positive effects of vaccinations, although it is widely acknowledged scientifically that vaccinations are safe, that they have been shown to be effective, and the potential for adverse effects is small (Boddice 2016; Dubé et al. 2016). Nevertheless, there is a growing disregard for these scientific research results, which has led to citizens acting in ways opposite to what said results would suggest: In Germany, for example, there was an increase in measles epidemics in 2015, which led to political debates about the necessity for compulsory vaccinations (Althaus and Salathé 2015; Betsch and Böhm 2015; Haverkate et al. 2012).

The case of climate change is similar: Some people simply do not believe the scientific warnings regarding climate change issues, or doubt the possibility of human-induced climate change (Matthews 2015). Beyond this scepticism it seems that there is a growing number of citizens who believe in conspiracy theories like "chemtrails" left by airplanes during civilian flights as part of giant geo-engineering activities (Bessi et al. 2015). Others believe that 9/11 has explanations other than those that have been officially stated (Krekó 2015). In social media, like Facebook, or Internet blogs, "alternative" theories or truths are distributed that sometimes receive a certain amount of trust (Matthews 2015). All in all, some argue that a new "war on science" is beginning, one that consists of neglecting scientific research, even in cases in which there is overwhelming scientific evidence available (Mooney 2006; Achenbach 2015). This "war on science" is fostered by actors with special interests who benefit from the uncertainties that normal people experience.

The reason for this seems to be that, in the past, science often failed to predict certain developments or consequences associated with technological innovations like atomic power, for example. This led to science having diminished credibility among the citizenry. In addition, in our modern world there are many information sources that can be used by citizens who cannot easily differentiate between trustworthy and biased or even wrong information. This is in line with growing scepticism regarding politics as a supporter of science-based solutions. These phenomena have led to the notion that a more effective communication between the public and science is necessary in order to counter this growing "war on science". The following example of more effective scientific communication on climate change issues shall demonstrate how these activities can be improved by the RIU model.

Evidence-based analysis by Kahan (2013) of the science-public communication revealed that the public conflicts over climate change are not caused by a deficit in public comprehension of scientific knowledge (Kahan 2013, p. 3). It is not citizens' irrationality that is driving the questioning, but rather an assessment of scientific information that is rational in regard to the "goal or interest" of specific citizens or groups they belong to. Even events that citizens perceive with their own senses are treated selectively according to the measure in which they are aligned with their own goals and interests. Furthermore, citizens who have better access to scientific results and understand them better have an even more likely to select specific scientific evidence with regard to their interests. Therefore, improving the quality of scientific information will not reduce its likelihood of being questioned, rather, it will increase it.

Kahan provides an excellent example for a selective assessment of science in the climate change debate (Kahan 2013). Citizens behave rationally if they try to align their beliefs regarding climate change with those of people who make up their peers. If, for example, they argue critically whether or not climate change threatens mankind, and, as a consequence of contrasting beliefs, this generates a conflict with family members, colleagues or friends, their quality of life will be affected quickly and seriously in a negative way. In contrast, consequences by climate change will be felt definitely not in the short run and remain vague for the individual citizen.

"People whose beliefs are at odds with those of the people with whom they share their basic cultural commitments risk being labelled as weird and obnoxious in the eyes of those on whom they depend for social and financial support" (Kahan 2012, p. 255).

These findings are the point of departure for the solution for effective science-public communication that RIU suggests. The main argument here is that there is a need to differentiate strictly between the realm of research and the realm of practice. Whereas in science scientific truth counts most, in practice it is the usefulness of the scientific results for actors that becomes most important. Between these lies integration, which optimises the coupling of scientific results and the interests of specific actors, who in this case are citizens.

The recommendations for efficient science-public communication differ with each of the three phases: research, integration, and utilization. A huge pool of sound scientific knowledge is needed as a basis for the whole process. It should reflect the most current knowledge regarding problems and their solutions. Of course, the only available scientific knowledge is often comprised of results expressed in terms of specific probabilities. Such knowledge can be better produced if research is as independent from the citizenry as possible.

In the integration phase, scientific knowledge meets the needs of citizens. A mutual selection process takes place in which some scientific results become part of the arguments and beliefs of specific groups of citizens. The selection criterion is —as Kahan's work revealed—how well the results match the interests and goals of those on whom the citizen in question depends for support. Integration cannot change the degree of dominance of particular interests, nor the dependence on the support from peers and others, but it can make the choices visible by showing alternatives solutions and the specific actor groups responsible for them. In addition, the scientists recommending specific alternatives become responsible for them and cannot delegate responsibility to science in general. To optimise the process it is important that the communication between researchers and citizens should be improved. What is most important is to keep in mind that greater input by researchers would not eliminate the different interests and forms of discourse within a given, shared point of view, but rather, it would widen the competing alternatives, in most cases. Nevertheless, a broad sample of scientific, sound alternative solutions strengthens the scientific rationality of any choice.

The final step in the RIU model is the utilization of the scientific knowledge in decision-making and in the choice and implementation of solutions. The RIU model argues vehemently that it is unreasonable to hope that scientific reasoning, or scientific "truth", alone will motivate citizens to accept and believe specific scientific results in practice. The causes for rejection are individual interests and dependencies. Therefore, a working solution that could change beliefs must deal with these interests and dependencies in a realistic manner. The RIU model describes it plainly: A sound scientific argument needs powerful allies who use their power to compel other citizens to accept the argument. Science is responsible for the truth but only actors have the potential to assert specific solutions in the political process. Democracy legitimates specific actors for such decisions. The RIU model

RIU Criteria		Effective Science-Public Communication
Independent science	Consensus/ contradiction	■ State of art in facts and probabilities about problems and solutions  **Responsible** Researchers and research organizations
	Probability	
	Unknown	
Pluralistic integration	Alternatives without agreement	■ Alternatives in problem solutions  **Responsible** Public alliances of scientists and citizen
	Link to actors and interests	
Democratic decision	Joint results	■ Decision in innovative solutions 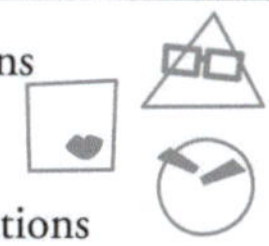 **Responsible** Multilevel - democratic institutions
	Joint responsible actors	

Fig. 5.7 Effective science-public communication according to RIU

recommends that democratic institutions should make use of their duty, right and power to make decisions. Even the best communication cannot produce a consensus among the citizenry by referring to purely scientific arguments to change discourse antagonism to harmony, and thereby substitute political decision-making.

According to the RIU model, effective communication between science and the public can be built upon sound independent research that provides full information to various alliances of scientists and citizen groups, who may believe in and promote different and sometimes competing alternatives in public. The wide range of alternatives supported by different groups of citizens does not indicate a failure of said communication. Rather, it shows the realms in which political decisions are needed (Fig. 5.7).

References

Achenbach, J. (2015). Why do many reasonable people doubt science? *National Geographic, 14* (5). http://ngm.nationalgeographic.com/2015/03/science-doubters/achenbach-text.

Althaus, C. L., & Salathé, M. (2015). Measles vaccination coverage and cases among vaccinated persons. *Emerging Infectious Diseases, 21*(8), 1480.

Banchoff, T. (2002). Institutions, inertia and european union research policy. *JCMS: Journal of Common Market Studies, 40*(1), 1–21. doi:10.1111/1468-5965.00341.

Barlösius, E. (2009). „Forschen mit Gespür für politische Umsetzung "–Position, interne Strukturierung und Nomos der Ressortforschung. der moderne staat–Zeitschrift für Public Policy, Recht und Management, 2(2), 347–367.

Barlösius, E. (2010). Ressortforschung. In D. Simon, A. Knie, & S. Hornbostel (Eds.), *Handbuch Wissenschaftspolitik* (pp. 377–389). Wiesbaden: Springer.
Beck, S. (2010). Vertrauen geschmolzen? Zur Glaubwürdigkeit der Klimaforschung. *Aus Politik und Zeitgeschichte, 32*, 15–21.
Bessi, A., Coletto, M., Davidescu, G. A., Scala, A., Caldarelli, G., & Quattrociocchi, W. (2015). Science vs conspiracy: Collective narratives in the age of misinformation. *PLoS ONE, 10*(2), e0118093.
Betsch, C., & Böhm, R. (2015). Detrimental effects of introducing partial compulsory vaccination: Experimental evidence. *The European Journal of Public Health*, ckv154.
Böcher, M. (2010). Federal Environment Agency (Germany). In D. Guston & J. G. Golson (Eds.), *Encyclopedia of nanoscience and society* (pp. 234–235). Thousand Oaks: SAGE.
Böcher, M. (2012). Wissenschaftsbasierte Politikberatung auf Abruf – Zur Rolle der Ressortforschungseinrichtungen für Ministerien am Beispiel des BMU. der moderne Staat, 5 (2), 459–480.
Böcher, M., & Krott, M. (2010). Umsetzung des Konzepts einer modernen Ressortforschung im Geschäftsbereich des BMU (Vol. 39/2010, Umweltbundesamt-Texte). Dessau-Roßlau: Umweltbundesamt.
Böcher, M., & Töller, A. E. (2012). *Umweltpolitik in Deutschland: eine politikfeldanalytische Einführung* (Vol. 50). Wiesbaden: VS Springer-Verlag.
Boddice, R. (2016). Vaccination, fear and historical relevance. *History Compass, 14*(2), 71–78. doi:10.1111/hic3.12297.
Döhler, M. (2007). Die politische Steuerung der Verwaltung. Eine empirische Studie über politisch-administrative Interaktionen auf der Bundesebene. (Vol. 11, Staatslehre und politische Verwaltung). Baden-Baden: Nomos.
Dubé, E., Vivion, M., Sauvageau, C., Gagneur, A., Gagnon, R., & Guay, M. (2016). "Nature does things well, why should we interfere?" Vaccine hesitancy among mothers. *Qualitative Health Research, 26*(3), 411–425.
Edenhofer, O., & Minx, J. (2014). Mapmakers and navigators, facts and values. [10.1126/science. 1255998]. *Science, 345*(6192), 37–38.
European Commission. (2013). *Options for strengthening responsible research and innovation*. Luxembourg: Publications Office of the European Union.
European Forest Institute. (2003). Convention on the European Forest Institute. Joensuu.
European Forest Institute. (2013). *Annual Report 2013*. Joensuu: EFI.
European Forest Institute (n.d.). Science Supporting Policymaking.
Falk, T. M., Rocha, S., Warnick, B. R. (2009). Social science and its discontents: An essay review of Bent Flyvbjerg's making social science matter. *Education Review, 12*(4).
Flyvbjerg, B. (2001). Making social science matter: Why social inquiry fails and how it can succeed again. Cambridge university press.
Flyvbjerg, B., Landman, T., Schram, S. (Eds.). (2012). *Real social science: Applied phronesis*. Cambridge: Cambridge University Press.
Geden, O. (2015). Policy: Climate advisers must maintain integrity. *Nature, 521*(7550), 27–28.
Geoghegan-Quinn, M. (2012). The dialogue between science and the rest of society has never been more important. In *Science in Dialogue Conference—Towards a European Model for Responsible Research and Innovation*, Odense, 2012.
Geuna, A. (2001). The changing rationale for European university research funding: are there negative unintended consequences? *Journal of economic issues, 35*(3), 607–632.
Haverkate, M., D'Ancona, F., Giambi, C., Johansen, K., Lopalco, P. L., Cozza, V., et al. (2012). Mandatory and recommended vaccination in the EU, Iceland and Norway: results of the VENICE 2010 survey on the ways of implementing national vaccination programmes. *Euro Surveill, 17*(22):pii = 20183.
Hey, C. (2009). Wirkungs- und Leistungsbilanz. In H.-J. Koch & C. Hey (Eds.), *Zwischen Wissenschaft und Politik, 35 Jahre Gutachten des Sachverständigenrates für Umweltfragen* (pp. 161–279). Berlin: Erich Schmidt-Verlag.

Hohn, H.-W., & Schimank, U. (1990). Konflikte und Gleichgewichte im Forschungssystem: Akteurkonstellationen und Entwicklungspfade in der staatlich finanzierten außeruniversitären Forschung (Schriften des Max-Planck-Instituts für Gesellschaftsforschung Köln; 7). Frankfurt/Main [u.a.]: Campus-Verl.

Kahan, D. (2012). Why we are poles apart on climate change. *Nature, 488*(7411), 255.

Kahan, D. M. (2013, February 13). Making climate-science communication evidence-based—all the way down . In M. Boykoff, & D. Crow (Eds.), *Culture, Politics and Climate Change* (eds. M. Boykoff & D. Crow, Routledge Press, 2014 Forthcoming).

Kraft, P. W., Lodge, M., & Taber, C. S. (2015). Why people "Don't Trust the Evidence": Motivated reasoning and scientific beliefs. *The ANNALS of the American Academy of Political and Social Science, 658*(1), 121–133. doi:10.1177/0002716214554758.

Krekó, P. (2015). Conspiracy theory as collective motivated cognition. In M. Bilewicz, A. Cichocka, & W. Soral (Eds.), *The Psychology of conspiracy* (pp. 62–76). Hove, New York: Routledge.

Krott, M. (2008). Forest Government and Forest governance within a Europe in change. In L. Cesaro, P. Gatto, & D. Pettenella (Eds.), *The Multifunctional Role of Forests—Policies, Methods and Case Studies* (Vol. 55, pp. 13–25, EFI Proceedings).

Laitin, D. D. (2003). The perestroikan challenge to social science. *Politics & Society, 31*(1), 163–184. doi:10.1177/0032329202250167.

Matthews, P. (2015). Why are people skeptical about climate change? Some insights from blog comments. *Environmental Communication, 9*(2), 153–168.

Mooney, C. (2006). *The Republican war on science*. New York: Basic Books.

Nature Geosci. (2013). Deja vu on climate change. [Editorial]. *Nature Geoscience, 6*(10), 801–801. doi:10.1038/ngeo1983.

Owen, R., Macnaghten, P., & Stilgoe, J. (2012). Responsible research and innovation: From science in society to science for society, with society. *Science and public policy, 39*(6), 751–760.

Pülzl, H., Hogl, K., Kleinschmit, D., Wydra, D., Arts, B., Mayer, P., et al. (Eds.). (2013). European Forest Governance: Issues at Stake and the Way Forward (What Science Can Tell Us 2) Joensuu: European Forest Institute.

Rose-Ackerman, S. (1995). *Controlling environmental policy: The limits of public law in Germany and the United States*. New Haven, CT: Yale University Press.

Schram, S. (2012). Phronetic social science: An idea whose time has come. In B. Flyvbjerg, T. Landmann, & S. Schram (Eds.), *Real Social Science: Applied Phronesis* (pp. 15–26). Cambridge: Cambridge University Press.

Stocker, T. F. (2013). Adapting the assessments. *Nature Geoscience, 6*(1), 7–8. doi:10.1038/ngeo1678

Umweltbundesamt. (2005). Ratgeber: Hilfe! Schimmel im Haus. Dessau-Roßlau: Umweltbundesamt (UBA).

von Storch, H. (2014). Theses on "Rethinking EU leadership in fighting climate change". http://klimazwiebel.blogspot.de/2014/08/theses-on-rethinking-eu-leadership-in.html

Warren, M. E. (1999). What is political? *Journal of Theoretical Politics, 11*(2), 207–231. doi:10.1177/0951692899011002004.

Wissenschaftsrat. (2007). Empfehlungen zur Rolle und künftigen Entwicklung der Bundeseinrichtungen mit FuE-Aufgaben. Berlin: WR.

Wissenschaftsrat. (2010). Empfehlungen zur Profilierung der Einrichtungen mit Ressortforschungsaufgaben des Bundes Lübeck: WR.

MIX
Papier aus verantwortungsvollen Quellen
Paper from responsible sources
FSC® C105338

If you have any concerns about our products,
you can contact us on
ProductSafety@springernature.com

In case Publisher is established outside the EU,
the EU authorized representative is:
Springer Nature Customer Service Center GmbH
Europaplatz 3, 69115 Heidelberg, Germany

Printed by Libri Plureos GmbH
in Hamburg, Germany